Michael Stocker

Hear Where We Are

Sound, Ecology, and Sense of Place

 Springer

Michael Stocker
Lagunitas, CA, USA

ISBN 978-1-4614-7284-1   ISBN 978-1-4614-7285-8 (eBook)
DOI 10.1007/978-1-4614-7285-8
Springer New York Heidelberg Dordrecht London

Library of Congress Control Number: 2013944294

Cover illustration: Cover photo of dolphins hunting by Alexander Safonov

Springer is part of Springer Science+Business Media (www.springer.com)

Hear Where We Are

*In memory of Henry B. Refo
and Leigh St. Maur Stocker*

In praise of *Hear Where We Are*

"I read your manuscript and was immediately struck by the importance of this work."
Gordon Hempton, Author, *One Square Inch of Silence*

"Michael Stocker has unique insights and encompassing information of sound in environments ranging from ocean floors to Gothic cathedrals. I know of no others who span such a wide variety of places with such a clear vision of how sound and acoustics makes places what they are."
Peter Warshall, Editor, *Whole Earth*

"What would the world be like as perceived by another animal? Locked into the limits of our own senses, it is difficult for humans to study or even imagine. *Here Where We Are* gives you a window into this fascinating subject. From the lateral lines of fish, to the whiskers of seals, to the acoustic light of snapping shrimp, animals "hear" in ways you would never suspect. It is exciting to imagine worlds of perceptions just outside of our own. I have never thought of the music and animal song that emanates from a tropical reef. With the right ears it would be as loud and wonderful as a tropical rainforest! As you read, you will find yourself getting down on the floor with one ear plug to discover ground level sound and dreaming of whale necropsy. Michael Stocker is a true naturalist and *Hear Where We Are* will open your mind (and ears) to a new world."
John Muir Laws, Naturalist, Illustrator, and Author,
The Laws Field Guide to the Sierra Nevada and *Sierra Birds*

"Michael has an excellent handle on the world of sound and acoustics and is able to convey his understanding to a broad audience. He is making a considerable contribution to the general understanding of how sound and acoustics work in the ocean environment."
Vivienne Verdon-Roe, Filmmaker and Academy Award winner for the film,
Women for America, for the World

"Michael Stocker's work has been enormously helpful to me in understanding the world we live in. Michael has an encyclopedic grasp of science, sound and technology which he is able to communicate in an easy and enjoyable way."

Hallie Iglehart Austen, Author, *Womanspirit* and *The Heart of the Goddess*

"With *Hear Where We Are*, Michael Stocker generously opens the gates into the vast and oft-mysterious world of human and animal acoustics. While his science is solid—and indeed startling in its breadth and ambition—it is his stories and deep empathy that are most compelling here. Over the course of five well-considered chapters, we are drawn into Stocker's world, a place where sound is no longer merely a means of communication, but takes its rightful place as a physical, palpable web of connection that permeates all life on this singing planet. His dual focus on human and animal sound-making leaves us with a much richer, and far more authentic, awareness of the ways that all species share a fundamental experience of sound, yet find incredibly diverse expressions of this primal thread of connection and creation. First, there was the sound. And now, thanks to this gem of a book, we can hear it anew."

Jim Cummings, Executive Director, Acoustic Ecology Institute

Acknowledgements

I was once told that writing a book is much like having a child; sometimes it is intentional, other times it comes about as a surprise. This one was a bit of a surprise. It was quite some time ago that I decided to write an article about the acoustical design of public spaces. I was then designing exhibit spaces for museums. While speech intelligibility and noise control are the rudiments of architectural acoustics, museum spaces particularly need to attend to the visitor's experience; conveying information, framing each individual exhibit and promoting personal autonomy while encouraging a family and community context. These design aspects all play into the visitor's emotions and subconscious "sense of place."

I described some of these characteristics of exhibit design in the article[1] and sent it around to a number of trade publications. Of the ten I sent it out to, seven wanted to publish it. Previously not thinking myself much more than a writer of contracts, the trade editors were telling me that I could write something worth reading. I decided to write a more detailed article looking at how architectural acoustics influences our emotions, and how architectural design can support or sabotage the purpose of inhabited spaces.

Three days and 50 pages later I realized I probably had a book in me and began pulling it out and laying it down on paper (or in Word files to be exact). Like any child it began by being a bit disruptive but relatively easy to control. I realized that in order to convey ideas about how we experience sound in spaces I needed to convey something about the experience of sound; how it can encourage a sense of safety and belonging, or induce a sense of dread and isolation. This led to an examination of how humans—and then other animals use sound to establish their participation in their own soundscape, and in the soundscapes of others.

Soon enough the child-book became the adolescent-book. It wanted autonomy. It did not want to share the covers with chapters on sacred spaces or the "sound of the commons." After trying really hard to reconcile these differences in opinion I acquiesced, dropping all of the work on architectural acoustics and let the book become what it is now.

Like any child, a book cannot be raised alone. In the beginning I was cataloging the names of the people who in some measure had helped. From the security guards at public buildings who were kind enough to let me explore their keep, or the Cistercian nun whose humility illuminated for me the sacred sound of the Rose; from bioacousticians such as Art Popper, David Mann, Bob Dooling, and Darlene Ketten for their feedback and input on animal hearing, to indigenous ceremonial elders Rupert Encinas, Luciano Perez, and Fred Wahpepah for framing perception in completely different dimensions. Both prison wardens and university professors had a hand or lent an ear. A conversation with Ivan Illich expanded my thinking on the cultural phenomenology of sound, and Gus the cricket generously unfolded his wings to reveal the secrets of his ventriloquism.

My list began pushing out at over a thousand kind and generous souls. I could write a book about these encounters alone—which is sort of what I did, and I owe them all a debt of gratitude for their contributions. But above all I want to acknowledge my father Godfrey Hugh Stocker who handed down to me—either through genetic predisposition or by example, the inquisitive mind of a generalist, and my mother Marion Lockwood Stocker for her "greatest work" of raising her own children by feeding their curiosity with all of the feathers and bones, salamanders and fish, any book we could digest—and letting us run free in the mountains armed with only a canteen, a pocket knife, and a wrist watch. I wish they were here to share this with in person.

Lagunitas, CA, USA   Michael Stocker

Contents

Introduction by Way of a Map

When was the last time you got a thrill hearing the rumble of distant thunder—just at the start of a patter of raindrops overhead? Have you ever been really pleased to hear the sound of someone's keys jingling at your front door? Did you ever notice your immediate relaxation in the silence of turning your office computer off? These scenarios are common experiences many of us have of our environment. We don't usually think of these experiences as "sound" because they are imbedded in other elements of our experience—the promise of a spring rain; the arrival of a loved one; the quitting of work for the day. But even when we hear these sounds on their own—separated from the events that produce them, they still produce compelling responses; invoking pictures in our mind's eye, emotions, and sensations in our bodies.

Hearing, like all of our perceptions, has adapted and developed to the level of complexity it has in humans for many reasons. While we largely frame sound perception in the context of communication and music, these are only two conscious aspects of sound and hearing. Equally important, and less apparent to us is that our hearing is a survival tool; it is a perception that allows us to perceive our environment in dimensions obscured from our vision, out of reach from our touch, and downwind from our sense of smell. Hearing is the perception that allows us to gauge the size, shape, and density of our surroundings, and sense our placement within it.

The overarching premise of this book is that sound perception is a perception of mutual engagement. It is the visceral bridge between us and others, and our contiguous connection to our surroundings. Sound takes time to unfold, and space to manifest. Our perception of sound blends these dimensions into physical sensations that we both affect and are affected by. And while our experience with sound unfolds as a "continuous now," our sound perception also invokes our conscious and subconscious past, and invites a lyrical—or fearful—anticipation of what is to come.

Even when we are not focused on sound—not actively listening, sound plays at the perimeter of our awareness; warning us, pulling at our attentions, informing us about the dimensions and dynamic conditions of our environment; allowing us to keep tabs on our horizon—all without asking to be monitored. And while sound tickles the perimeters of our consciousness, our sense of hearing also resides in the center of our sense of location and placement; our sense of balance, momentum, vector, and velocity are seated in the organs of the inner ear. Even our sense of unfolding time—melody, rhythm, and tempo—is informed by auditory stimulus. In sum, our physical sense of "place" and belonging is enhanced by the continuity of sound as it embellishes our perception of gravity and movement, mass and density, permeability, proximity, time, change, and physical vibration.

In the persuasive immediacy of our visual culture, we tend to forget that our visual surroundings are imbedded in a vibrating field of acoustical energy—a field that doesn't leave us when we close our eyes. This book explores this vibrating acoustical field as a fabric within which our living experience is woven; our communication, our sensations, our emotions, thought, and will. This visceral connection to our surroundings serves both as a reaching out and a collecting within. Our active expressions and our passive experiences are all at play in a resonating, vibrating, and sensuous soundfield. Because this soundfield envelopes the realms of all of our endeavors, the exploration in this book will by course intersect a broad range of experiential phenomena: Cognitive and subconscious, human and other animals' intellect, instincts, and behaviors.

While sound perception is not exclusive to humans, the first two chapters focus on human experience—in terms of how we are affected and influenced by sound, and how we use sound to affect and influence our surroundings. In the first chapter we examine how the elements of the soundfield within which we reside—the noises, familiar voices, the alarms, the sounds of media and technology, and the acoustics of our surroundings—influence how we feel. Our sense of power, size, confidence, and security are all impinged on by the sounds around us. In this chapter we find that our feelings of courage, anxiety, alienation, and conviviality are as seated in our acoustical surroundings as they are influenced by our social (or antisocial) settings.

If we were just passive experiential beings, we would be buffeted by these dynamic soundscapes. Fortunately, humans are also provided with the equipment to affect our acoustical surroundings and the wherewithal to cocreate our soundfields. Chapter 2 explores how we craft and modify our soundfields to serve our intentions. From seduction to warfare, from music to noise, from suspicious surveillance to compassionate healing; humans (and other animals) have consciously used sound to set our boundaries and perimeters; cultivating inclusion with those who are wanted within, establishing exclusion for those who are not, and teasing or testing others across that imaginal line.

Of course none of this would be possible without acoustical energy and the organs to perceive it. Chapter 3 sets out to unfold the phenomena of acoustical energy through the physical experience of sound. In this chapter the mechanics of sound are revealed in the context of our physical surroundings, and the way that acoustical energy impinges on our bodies, our skin, chest, guts, and especially our ears—manifesting it into the neural impulses that we perceive as sound.

This auditory transaction is necessarily very complex; conferring the entire acoustical signature of our environment into an extremely precise, finely detailed perceptual sound-scene—a scene upon which our very survival hinges. Within this scene we are in balance while at rest, though poised for immediate flight at any instant; relaxed in familiar surroundings but ever aware of threat or opportunity just outside that known realm.

In the preponderance of the literature, our "organs of hearing" have been deconstructed into constituent parts and examined as a set of mechanical transactions: The ear drum, the auditory ossicles, the cochlea, the semi-circular canals—then reassembled into various models called "the ear." We know that the models are lacking because there is more than one version, and medical technologists are just beginning to agree on which is the most useful (and under what circumstances). And still these models only approximate the acuity of our hearing. Scientists are still hypothesizing how we locate sound sources on our neural map with such refined accuracy; how we are able to discriminate delicate sounds buried in loud noise, and how a very simple sound can trigger a flood of emotions and thoughts.

Not only are the shortcomings of the models a consequence of the vast complexity of our hearing process, but they are also a result of systematically ignoring some of the fundamental forms of the ear. These forms are necessarily complex because the hearing process translates our entire auditory world—from sound to sensation, through an exceedingly small envelope and with phenomenal precision. The third chapter orbits around these fantastic forms from a fresh perspective, teasing a bit more out of the mystery of sound perception.

Human hearing is so fabulous and so well adapted to our needs that it is easy to believe that it represents the apogee of bio-acoustic adaptation. It has been from this perspective that most studies of animal sound perception have been filtered—using human perceptual priorities as benchmarks. But on close examination it is clear that each animal in its own habitat has adapted to its acoustical setting in ways most suitable to its own priorities. These adaptations are often outside of our human ability to fully comprehend. Various animals live in realms where pitch discrimination and amplitude sensitivity give way to subtle phase and time-domain distinctions; where localization cues are derived through substrate vibration and where serial streams of sound give way to spatial constructions in four dimensions.

Chapter 4 examines these complexities in some of the many worlds of animal bio-acoustics; from sea to shore, from savannah to sky, from spider webs to subterranean warrens. Each acoustical niche presents sound in unique ways—all suitably reflected in the myriad of adaptations found in the resident creatures. Bearing in mind that these creatures include the hunter and the hunted, the sedentary and the mobile, the land-bound and the free-floating or free-flying, the range of adaptation is vast—all perceiving acoustical energy in ways that are largely outside of our perceptual grasp. Exploring these diverse perceptual realms gives us an opportunity to appreciate the myriad dimensions of sound experienced by the cohabitants of our world, and thus expanding our own experience of sound.

With a diversity of perceptual tools in play, the idea of "communication" is set in motion: The interplay of body and space, memory and response, form and gesture— all constituting sets of interactions between individuals and kin, kin and community, and organism with environment. Communication is so much more than groups of representative words set in a tapestry of grammatical rules; rather it encompasses how we establish our relationship with our surroundings and with those with whom we inhabit it. Chapter 5 settles into the foundation of "sound communication" as a visceral connection with others through sound; mediating the ethereal realms of acoustics with the tangible realms of space and consequence.

The larger mission of this book is to lead us back into a stronger appreciation of the world in which we live. I hope that the journey through this work will enhance the reader's perceptions; inviting a deeper listening, and ultimately encouraging a more conscious participation with their environment—allowing you, the reader, to truly hear where you are.

A Note on Endnotes

I have included a reasonably thick body of endnotes to support the assertions I make throughout the text. I use them in a number of ways, and I hope that the reader finds them informative. The field of sound perception is so vast, intersecting so many disciplines, that any general work on the subject is bound to be incomplete to gossamer extents. For this reason I include endnotes that may serve as doorways to deeper inquiry—and if the reader finds a particular topic intriguing, the endnotes will hopefully provide some references to more complete writing on the subject.

Hear Here: The impact of sound on personal placement

> *"She quickly bounced to her feet with a chorus of jingles and chimes and started down the hallway." Don't you just love jingles and chimes? I do," she answered quickly. "Besides, they're very convenient, for I'm always getting lost in this big fortress, and all I have to do is listen for them and I know exactly where I am."*
>
> (Norton Juster, "The Phantom Tollbooth," 1961, Random House, New York, p. 145)

As I sit here and soak in these waters, I listen to the silence that surrounds me. Lightly punctuated by a drip of condensation from above, or a bubble releasing from below, the quietude of a Sierra hot spring saturates me to the bone; the warmth of the water, the pulse of my heart, the absence of sound. It is late at night—perhaps 2 a.m. The heat from yesterday's sun, rising up from the earth, has momentarily reached equilibrium with a gentle mist suspended beneath a low cloud cover. The night pauses, there is no breeze, it is dead silent.

As if to test this silence, I hear the choked cough of a red fox coming up from a close ravine. A cricket tentatively chirrups once… a few times… an owl inquires, a pebble falls into a ditch. Softly the night has shifted; the cool midnight air delicately exhales into the treetops of the surrounding ridgelines. This breath moves in and around slowly but purposefully, gently stirring the rushes in a nearby meadow, bringing on the slow seep of the dawn.

I come to these springs for the waters, but I also come for the silence. Though when I close my eyes it is not silence that I hear… just less sound… sounds I can almost count and identify; water, air, rock; the scraping of a beetle on a branch and the flutter of a moth against the sconce of a flickering candle.

When I leave the pools to sleep the alpine breeze whispers into my dreams as it mixes midnight into morning. I am stirred briefly awake by the first strains of the dawn chorus and the waking of the birds, returning to my slumber to be finally awakened by the murmur of human voices at a nearby camp; the sound of fire and cooking, metal pans and flatware, the cracking of eggs and the sputtering of sausage. The day has arrived and I am in it…

We are always submerged in sound and vibration; it excites our ears and touches our bodies, our skin and our bones. Sound perception is not voluntary. While we can turn off the lights or shut our eyes to hide our surroundings from view, we can't as easily shut our ears—we do not have "ear-lids." If we occlude sound from our thoughts, it is only meaning that we lose—the sounds that convey the message still strikes our ears, resonates in our chests and glances off our face.

We usually consider sound in terms of the "useful" sounds we can qualify, such as the sounds of music and language, but subconsciously we are more dependent on the incidental sounds of our environment to reveal the hidden dimensions of our reality. The sounds of our surroundings enable us to gauge where we are, how safe or exposed we feel, and our position of dominance or deference in our social and spatial settings. The feelings induced by these sounds influence our ability to act; they inform our communication and affect our willingness to speak. Our response to these feelings is often to make sound—testing our autonomy against our surroundings. We are assured when we hear the sound of our own footfalls; we may be emboldened if our voice or our footsteps robustly fill our environment or may feel timid if those sounds are not quite large enough for our surroundings. As we listen for meaning and information, we are also subconsciously probing our environment through the sounds we make. A cough, a deep breath or a gentle sigh will give us running affirmations of our location and the impact we have on it. The whir of a computer, the clatter of a restaurant, or the thunder of freeway traffic will frame our location and influence the feeling of autonomy that we have in it. We rely on visual cues to confirm our whereabouts; we collect information through our eyes, but our experience of where we are is often more dependent on sound.

Try this experiment: After reading the following paragraph, close your eyes and listen to your surroundings—try to identify what you hear. Are there birds, cars, or dogs present? Is your neighbor mowing her lawn? Is someone talking on the phone nearby, are there airplanes overhead? Do you hear water running or wind in the trees? Is your refrigerator humming? What are the elements in the environment outside of your field of vision that tell you where you are? What or who is outside of your realm, but still in your world?

This little "hear where you are" exercise is something I do when I feel jangled from "information poisoning"—being subjected to too much stimulus. It allows me to gauge myself in my surroundings solely through my experience of sound—I use sound to help me feel "in place." If I am indoors, I can hear the size and texture of the room I'm in. I can hear if the room has concrete, wooden or carpeted floors; whether I am in a small home or a large building, and if I am in the city or in the country. I can verify my placement in the room by the sound of my chair squeaking and the air whispering through my nose and lips as I inhale and exhale. I can affirm my location with my voice, in a sigh, a cough, a call, or a song. I may do this consciously and intentionally, but even when I am not deliberately exploring the acoustics of where I am, the sounds that support my experience of place continue to surround and influence me.

The common understanding is that we humans generate sounds most specifically to convey meaning in the form of language, but much of the sound and noise we generate conveys little new meaning and may be more driven by our need to constantly gauge ourselves within our surroundings. The noun "sound" refers to the acoustical energy that we hear, but the verb "to sound" refers to an examination, to find out, question, or query. The allusion is to plumbing the depths of the ocean, to gauge a dimension which we cannot see. The maritime principal of "sounding" is still used in lieu of radar for guiding ships through narrow channels and away from rocks in foggy weather. Ship pilots use sound to find out where they are in their ocean and surrounding geography just as we use sound to place ourselves in our own terrestrial surroundings. We may not be consciously aware of it, but we use this principal of "sounding" continuously as we make and perceive sounds. Sounding helps us gauge the socially appropriate volume to use when we are speaking to others; if we feel confident, we may deliver our strident speech in a voice more than adequate for our message; if we feel fearful, we may retract our voice to a whimper—or in attempt to embolden our fear, we may shout—as if in anger—to push back what is making us fearful. We set our will against our surroundings with the sounds we make, but even in silence the sounds around us are constantly affecting our emotions. The world enters us by way of sound—we sound back to gauge where we are in the world.

In this chapter we will examine the subjective experience of sound and how it plays into our emotions and feelings. Our exploration will transect the experience of sounds that warn us of danger, keep us alert, invite our comfort, or lull us into calm. We will find that our individual responses to particular sounds may differ, but they are framed in the biological responses to sound cues that we have in common with other people, and even other animals.

Sound and the Perceptual Body

Our sense of personal placement through sound perception is stimulated by the fact that sound reaches us through our bodies as well as through our ears. It is easy to localize sound perception to the organs of hearing—our ears—but rapid changes in pressure gradients over time in our environment—what we call "acoustical energy"—stimulate our bodies as well. We feel sound in our chest cavities and against the broad plain of muscle and skin of our backs. The nerves surrounding our hair follicles, the positioning sensors of our skeleton and the delicate nerves on our cheeks are all sensorineural pathways that stimulate auditory processing in our brains. Our sense of sound includes the embrace of our body by the environment.

Determining whether sound perception through our ears has precedence over the sensation of sound through our bodies is perhaps moot; at any given moment we may be consciously focusing on hearing sound while simultaneously responding to any degree on how our body is stimulated by sound in our environment. This distinction may be revealed in how well our body inhabits space. We may be agile and responsive, dancing through our surroundings at one time, and clumsy and uncoordinated, stumbling and colliding with things at another. At any time we may find ourselves tuned into the music of our surroundings, or oblivious to the rhythms and noise we inhabit.

Our personal sense of placement within an environment depends on two distinct realms of hearing; our sensation of the sounds we personally generate, and our perception of the soundscapes we are within. This sense of personal placement is substantiated by the fact that our perception of sound does not reside exclusively within our ears—it includes our entire body.

By way of example, the seat of confidence is in our heart; the heart is at the core of the word "courage," which we express through the motions of placing our hand over our heart, or striking our chests when declaring our convictions. The strength of our voices resonating within our chest cavity helps set our confidence; the relationship between the strength of this resonance and the strength of what we hear reflected back to us from our surroundings gives us an impression of how subjectively "large" we are. If we are in an intimate, close environment (such as a breakfast nook) we don't need to expand our voice to fill it, so we speak on a comfortable *sotto vocce*; if we are in a huge voluminous reverberant space (such as a cathedral,) we may fear meeting its size, compelling us to whisper. These responses have less to do with size of a space than with its acoustical setting. A shower stall or a large auditorium—dimensional equivalents to the breakfast nook and cathedral examples—may compel us to sing out or yell rather than hide our voice. The differences largely reside in how we hear ourselves in these surroundings. Through sound, we can gauge the power and size of our environment, our personal power and size within it, and our ability to transform it through our personal power. The extents of our self confidence are continuously at play on the perimeters of our perception of sound.

Cultural and Gender Distinctions

Within this context, everybody inhabits a distinctly individual soundscape, dynamically responding to our surroundings and to others that inhabit it with us. Personalities notwithstanding, our individual sense of sound perception is also influenced by social, cultural and even economic meta-factors that establish the backdrop of our auditory sense of who and where we are. One clear example of this resides in the perceptual artifacts of human sexual dimorphism—the distinct physical traits that define gender and also engender perceptual differences between men and women.[1] These gender-based perceptual adaptations influence how each man or woman responds to their surroundings. In the archetypal "hunter/gatherer" social structure this is represented both in terms of the gender roles in the community as well as predictable gender responses to stimulus.

A common framing about historic hunter/gatherer community settings is that the chatting, singing, and gossip of women's work circles was in part a strategy used to keep their community sound alive within an otherwise hostile environment. When the men were off silently hunting in groups, the women—who were somewhat unprotected, would both ward off predators and keep aural tabs on their community by generating a constant flow of sound. The stopping of sound within the women's circle would trigger fear and signal danger—some threat from outside the circle, perhaps even the abduction of a community member. Conversely, when out stalking their prey or protecting the perimeter of their camp from predators, the men needed to be silent. For the men, a sudden outburst of noise announced an encounter, triggering fear and signaling danger.

While the hunter/gatherer society is rapidly disappearing into the blur of globalization, many of the gender influenced perceptual artifacts remain. Women's desire to speak about, and communicate their emotions may have grown out of a historically accepted role for women to "keep the community fires burning" and to know what is happening within the circle. Meanwhile, men's traditional need for silence in unpredictable and changing environments may be borne out of their need to silently survey and assess their surroundings when they feel threatened. Somewhere between these two gender perspectives of sound and silence, safety and danger—men and women attempt to communicate.

Of course this line of conjecture is an over-simplification of our gender-acoustic responses to our soundscape, illustrating how creatures of the same species can have completely different relationships to sound and silence, hearing and listening. It does not account for the varied qualities of the sounds or the textures of the silence, or the distinctly different relationships that each individual has with sound.

There are no real blanket statements that generalize human responses to specific sounds; in the end, most responses to sounds and noises are learned through cultural context, environmental experience, and social setting. For example, some cultures have traditionally been considered "quiet and demure," while others "loud and boisterous," but these characteristics have as much to do with their surroundings as with their social predilections. Similarly, physical acuity in sound perception, once considered a "culturally" distinct attribute of some peoples—are more likely due to the acoustic environment of the specific groups rather than any cultural predisposition. For example both the Inuit of the North American continent and the Mabaan people of the Sudan are often referred to in audiology texts as being representative of tribes who have refined hearing because they've dwelled in a world of quietude. Comparisons of "normal" and aggravated hearing loss often used these people as benchmarks. The Inuit were once known for their quiet arctic hunting life; the Mabaan didn't have a drumming tradition and the murmur of their villages was referred to as "quieter than a refrigerator" in one text.[2] In their quietude, each of these groups also had very distinct responses to the sounds of their surroundings; the voices of the wind, the rustle of their unique fabrics, and the songs of their harvests would by nature and setting be very different from each other and from our own culture.

Today these "cultural" distinctions of sound perception are much more difficult to determine than they were 50 years ago because the sound of the dog-sled team for the Inuit, and the murmur of the Mabaan village have been replaced with sounds known across many cultures. For the Inuit families, the barking of a husky team bringing the hunters home—the warm herald of arrival and abundance, has been replaced by the whine of the snowmobile. The Mabaan, now caught up in a revolution, are carrying rocket launchers and AK-47's. And while 50 years ago the helicopter was not a recognizable soundmark in the Mabaan lexicon, it is likely that they now have the same adrenalin saturated response to the sounds of helicopters that Cambodian and Guatemalan villagers have (who have also learned these sounds through war and political strife). To the Mabaan, Cambodians, and Guatemalans, the sound of attack helicopters invokes the arrival of hopeless death and destruction.

There are some predictable responses to archetypal sounds across all cultures, such as the sound of water, fire, and heartbeats. (These sounds are archetypal due largely to how they play into our sense of survival.) The "fingernails on the blackboard" experience tends to affect most sentient beings in a similar manner, and other alarming noises—such as explosions and the sounds of falling things—typically stimulate adrenalin production and fear response in all who hear it. This may indicate a deeper predisposition to sounds of certain dimensions or qualities rather than sounds of particular things. Human babies, for example, are born with only two innate fears; the fear of falling, and the fear of loud noises.[3] Both of these fearful perceptions stimulate the auditory centers of the brain—the fear of falling through the sense of motion and balance residing in the inner ear, and loud noises by way of the mechanisms of hearing through the ears and body. Both of these fearful perceptions summon the distinct possibility of rapid change, and the possibility of time slipping out of our control—or perhaps more germane to this discussion, the possibility of losing our grasp on where we are.

Many of the sounds that predictably trigger fear contain deep, low frequencies. Thunder, earthquakes, explosions, roaring, and bellowing—all imply that the source of the sound is both larger than we are, and not in our personal control. Conversely, the sounds that predictably enliven and please us seem somehow in the range of our containment, thus the gurgling and cooing of infants, the song of birds, the gentle trickle of water and the whisper of a breeze are all somehow comforting. We are assured by these sounds as they open up our perceptual world containing us in what suggests a sense of peace, and a feeling of inclusion.

Our sonic relationship with our surroundings begins at a very early age. Nerve endings appear in the inner ear of a human fetus in its 12th week of development. While actual hearing responses do not become apparent until the 24th week, the inner ear becomes vital and functioning in the 7th week of development,[4] enabling the embryo to establish equilibrium and balance itself within the womb. Before the ear facilitates perception of the "outer" world, it is the organ which anchors placement within the prenatal human's dark, amniotic domain.

A fetus is not equipped to make sounds into its surrounding, but the womb is hardly a quiet place. The heartbeat, blood flow, digestion, and metabolism of the mother creates an ambient noise environment that can exceed 55–65 dB in sound pressure level[5] (dB-SPL)[6]—the noise level found in a busy restaurant. Add to this the wealth of sounds coming in from outside the mother's body and we find that the infant gets quite a preview of things to come through sound and hearing long before they are born.

Fig. 1.1 Madonna and child by Il Sassoferrato (Giovanni Battista Salvi) (1609–1685)

Beating hearts set the tempo of this prenatal life—a sound that will dance and resonate within us all our life. The aural bond with our mother's heart keeps us close to her as infants. Desmond Morris in his popular book *"The Naked Ape"* noted that human mothers, regardless of their "leading" hand, tend to hold their babies so that the child's head is over her left breast—and her heart.[7] He speculates that this creates an important affirmation of the living bond between mother and child, continuing the fundamental sonic imprint that began the child's relationship with all things "outside" and bonding them to where they are.

Pediatrician Dr. Brian Satt of "A Sound Beginning"[8] made a practice of teaching expectant parents how to take advantage of this early contact with their child by singing it "Womb Songs"—any song that the parents were comfortable singing to their prenatal baby. These songs, sung to a child in utero imprint on the child and can be sung to help the child feel safe and comforted throughout his or her early life.[9] Results from Dr. Satt's work indicate that any sound coming into the prenatal environment affects the child, and that sounds heard while in the womb will invoke emotional responses in any of us throughout our life. This could explain why the sound-absorbing acoustics of soft cozy places comforts many of us—being similar to the acoustical environment of the womb. It may also provide clues into why some people are claustrophobic; perhaps the womb did not feel safe for some reason, or there was a strong dissonance between the womb and the world into which they emerged.

From the first draft of breath, a child begins probing and affecting their surroundings with sound. While an infant may not be able to always get things exactly right with their sound, they know that their sound will invoke change. Their lexicon may be simple, but any parent will tell you that it is very effective. Being otherwise relatively immobile, the child uses sound to weave their lives into the fabric of their surroundings. The sounds they create influence their caretakers, the sounds they hear indicate action. The exchange is simple: "I make noise, you guys move…" As we grow older, we develop more complex relationships with our surroundings, but this fundamental premise of "sound action and effect" establishes our placement in the play of silence and noise in the soundscapes that we inhabit.

The sound we generate expands into our surroundings, allowing us to gauge our "size" and influence within it. Our willingness—or need—to test our environment is dependent on the sounds already at play in it. This inclination hinges on the acoustical setting and how the environment answers back.

When I was in my late teens I lived with a woman and her young child. When he became a toddler he would thunder around our neighborhood in his "Big Wheel" tricycle exploring his boundaries.[10] One morning I heard him yelling in the parking lot out behind our house. When I went out to investigate, what I found sort of busted me up. Apparently he had discovered the sound of his own voice echoing off of the back wall of an adjacent building. He was gleefully exploring this echo by hurling epithets at the wall and listening to them hurl back at him as the sound of his voice bounced off the distant wall. At a lack of anything appropriate to yell, of course he resorted to obscenities—which seemed to carry some erotic value for him as well, because with his diapers dropped around his ankles and his hand in his crotch, he screamed "Weenie! Weenie! Weenie!" "Diapers!" "Poopoo! Poopoo! Poopoo!" When he noticed that he had been caught he was so embarrassed, I don't believe I had ever seen him blush so red. All I could really do was acknowledge his incredible find and show him some other noises to make.

As the breadth or our experience increases, our ability to test our surroundings becomes more refined; and while some people will always have a need to be loud, many of the sound cues that confirm our relationship with our surroundings pull back into the subconscious mind. We keep tabs on our sense of place without consideration. A sigh, the sound of our own footsteps or the squeak of our chair is often enough to maintain our acoustical connection to our surroundings.

The sounds we commonly hear—familiar voices, the acoustical profiles of our homes and environs, the opening and closing of doors, the arrivals and departures of friends and kin, and the reoccurring soundmarks of time in our community—all become set into our perceptual map. This map frames our soundscape. It is by nature dynamic, anchoring our perception of time because it needs time to unfold, and as it does unfold we subconsciously incorporate and discard, consider and disregard the events occurring in it. We adapt our own sonic imprint by modulating the volume of our voice, the depth of our breath and fall of our feet. Our level of conscious consideration depends upon how familiar or unique our soundscape is. Our sense of comfort, anxiety, or alarm is affected by how the soundscape responds to our own sounds, or how well we can blend into it.

If there is one blanket statement that can be said about sound and how we are affected by it, it is that sound is a perception of inclusion. By gauging the size of our body through the sensations of the sounds we feel—both internally and externally, we establish our relationship to our surroundings; we connect to our environment. How we mediate this connection, consciously and unconsciously, deeply influences our sense of being—of who and where we are.

Our Soundscape: At Play Between Silence and Noise

Just as there is an infinite variety of sounds to perceive, there is also an equal or perhaps greater array of silences. We are submerged in sound, but our sonic world is surrounded by silence. Silence frames all sound that we hear and all of the sounds that we make. We reach into this silence with our own sounds to probe its breadth and plumb its depth; though after our sounds have faded, the silence remains—as varied as the color black, as rich and complex as the darkness of night.

How we feel in this silence depends on its density and texture, but it also depends on how the silence occurs. The expectant silence suspending in the moment that an orchestra conductor raises his baton is far different from the silence that occurs between the strike of lightning and the crash of thunder (though the conductor may not agree). The silence within a Gothic cathedral is very unlike the silence in a Native American sweat lodge, and hospital silence is very unlike library silence. There are silences signaling a beginning soon to be filled, and silences that set the repose of an ending; the silence at the apex of an inhale, and another at the nadir of an exhale; the silent vertigo after a military fire fight, and the cradling silence after making love; the dense, pregnant silences of life and expectation, and the cold, empty silence of death.

All of these silences both surround our feelings and saturate our emotions; they encompass all that we do, but they also dwell inside of us with a power that can make us feel small—and the larger the silence, the smaller we may feel. It is perhaps this axiomatic fact that compels us to generate much of the noise we subject ourselves to. Teenage boys with "sound blaster" music systems and roaring skateboards, drivers honking their horns as they drive through tunnels; the nerve-shattering screams of little girls at play, and defense engineers blowing up ever larger explosives—all contain manifestations of the noisemaker's desire to feel and hear their personal power against the silence that surrounds them. Generating their noise

Fig. 1.2 Quiet through tunnel! (photo by author)

allows them to feel empowered—creating acoustical territories—almost always to
the annoyance of those within the sphere of their sonic influence.[11]

Oddly, to the noisemaker, their sounds are usually not noise at all—because they
are involved in it. What constitutes the experience of "noise" is entirely subjective
and ambiguous—as it involves a sub-conscious to conscious continuum of noise
awareness from "un-noticed" to intolerable.

While we may initially qualify much of the sounds in our soundscape as "noise"
or unwanted sound, much of the soundmarks in our soundscape actually serve to
place us in a contiguous setting. We produce our own sounds to shape and affect our
surroundings, but there are other noisy beings and things out there that are also
doing the same. As I write this to the intimate whir of my computer, the noise of the
refrigerator insinuates itself into my soundscape. My neighbor is playing with his

daughters while a passenger jet tears a distant seam across the sky. My other neighbor is out in his yard transacting business on the phone. A sports car passes by my house.

All of these noises are intentional to some degree. None of these sounds are really annoying—it would be uncanny if they were not there. This is where I dwell. I have habituated to my typical auditory surroundings and I can gauge myself to them with unconscious ease.[12,13]

So the idea of "noise"—while subjectively understood by everyone, is much a matter of perspective. To me noise is anything that interferes with my ability to comfortably dwell in my surroundings—my ability to communicate and otherwise establish my presence with the sounds I make. Noise compromises my intentions and impinges on my personal and community boundaries. To the gentleman who rises each weekday morning at 7 a.m. to idle his Harley before blasting off to work, the sound of its reckless rumble makes him happy—it is music to his ears; I don't find it pleasant. To the industrialist, the thundering of machinery in his factory is comforting; to the laborer in the same factory, the noise may create tension—alertness to the danger of being crushed or maimed by the machine. The sound of crashing ocean waves, comforting to those living within reach of coastal waters would only be confusing to those living in the high plains of Patagonia or Katmandu.

The word "noise" shares common roots with "nausea"—seasickness, and from the Latin *"noxia"*—to hurt or injure. Nausea is caused by unpredictable or unpleasant motion sensed in our inner ears, so it is not surprising that noise and nausea both reside in the organs of hearing. Noise annoys us from outside, impinging on our sense of hearing, aggravating our mood; nausea dissettles us from within, aggravating our sense of balance in our inner ears and causing queasiness in our gut.

We typically define "noise" as unintentional or unwanted sound—sound that distracts or confuses our perception of the sounds we want to hear. But as we delve into the realms of hearing, we find the differences between wanted and unwanted sound less clear—blurred by the distinctions of conscious and unconscious sound perception. We discover that sounds we have never consciously considered are nonetheless very important elements in our familiar and comfortable soundscape. We also find sounds that we are intimately aware of sneaking into the back door of our consciousness, affecting us in profound and unpredictable ways. The idea that we habituate to our sonic environment, and thus "block out" useless noise is only partially true; it assumes that we have a static relationship to our familiar surroundings which we somehow integrate into our state of being. While we may not consciously interrogate the noise field within which we dwell, the noise influences our relative sense of comfort, anxiety, and even physical health.

This becomes apparent when we pause for a moment to take stock of how we are affected by the incidental sounds of our modern culture—from its industrial beginnings with the rumble and bang of forced combustion and steam compression engines, to the whir and hum our current media-driven society. Initially the sounds of technical and industrial progress were welcomed. The noise of nineteenth century industrial factories—like their smokestacks in the horizon—held a fond place in the hearts of the surrounding community.[14,15] Signifying economic stability and abundance, the din of industry was once the song of optimism. Asked if the noise was bothersome, a factory town resident might respond "what noise?"

Over the last century in industrialized nations, the sounds of physical power have evolved into the sounds of information power. The machines, now more efficient, don't bang and boom, rather they whir and hum. We still push their sonic artifacts into the backwaters of our conscious mind. With so much information to consider, we only select outstanding items to ponder; the bulk of our soundscape serves as a background wash—the noise over which specific recognized sonic events occur. Within this noise field our innate curiosity reaches out to derive meaning from the din; we create a subconscious counterpoint to connect the elements of our cacophonous world into a coherent whole. Once we recognize it as a coherent whole, we don't waste processing power reconsidering it. The noise has set the tone of our surroundings, so we leave it to temper our perceptual world largely unchecked—even while the noise and clutter aggravates our sense of comfort and stimulates anxiety.

Architect and lighting designer William M.C. Lam evaluates comfort criteria of environments by how they serve the biological needs of the users. To feel comfortable we need environmental information that supports our principal biological needs. The information we need has to do with location, protection, the presence of living things, territory, refuge, time, weather and opportunity. If elements of these needs are in flux, or if their qualities are ambiguous, we "…become uneasy and uncomfortable. If the incoming data are clear and if the perceived facts indicate that everything is as expected and under control, we relax."[16] By implication our comfort level is directly proportional to our ability to contain, identify or control the significant elements within our environment which may challenge our biological needs.

> We are comfortable when we are free to focus our attention on what we want or need to see, when the information we seek is clearly visible and confirms our desires and our expectations, and when the background does not compete for our attention in a distracting way… We are distracted and made uncomfortable when the visual information is irrelevant or confuses our understanding of the environment. Our discomfort is increased when visual noise—irrelevant or confusing signals—dominates the field of vision and interferes with the ability to perceive relevant, useful facts about the nature of the environment or the progress of activities.[17]

In order to identify distracting or confusing elements in our perceptual realms, we need to focus on them. Locating the sources of clutter or noise requires us to consciously ask "Is this environment cluttered or ambiguous?" This may be easier to do in the visual realm due to the fact that we are more accustomed to establishing sets and boundaries based on visual cues. But we don't build auditory sets in the same manner—even though the impact of noise and clutter on our sense of comfort is no less compelling. It may be even more significant due to the subconscious nature of our auditory panorama; we can be distracted by acoustical confusion and sonic clutter without consciously identifying it.

Sound is caused by, and thus associated with things in motion. With a higher level of cacophony comes a higher probability of something unexpected occurring, so we are naturally more alert in noisy environments. Even when the noise level becomes accepted and thus "blocked out," an underlying level of anxiety remains. This is probably a contributing factor to the significantly higher miscarriage rates of women who live in flight paths of airports.[18] It may also explain other physiological and neurological complaints found in people who live in and around high noise levels.[19,20]

These are just symptoms that connect environmental noise levels to states of health and well being. What we easily lose track of is how pervasive the problem is in urban and industrial settings, and the impact that environmental noise can have on our individual as well as collective survival. Psychologist Theodore Wachs' study of environment and childhood development indicated that noise in the early home environment is a strong factor in slowing down language and cognitive development.[21] Likewise, Dr. Arlene Bronzaft's study of learning abilities of elementary age students in a New York City school indicated that their reading skills were significantly compromised by noise from an elevated train track adjacent to the school. She found that students in classrooms facing the tracks were as much as a year behind their reading ability by their sixth year compared with students on the quieter side of the building. Once the Transit Authority installed acoustical treatment on the tracks, a follow up study showed no difference between the two groups.[22]

These quantitative studies infer deeper relationships to environmental noise; the literature is rife with studies on the affects of noise and aggression, noise and depression, noise and anxiety, noise and loneliness, etc.[23] The data in these studies are alarming, but the conclusions are not surprising. Environmental noise impinges on our senses, our bodies, our emotions and our reason—probably to a greater degree than we know. We can modulate the noise by constructing walls, attenuating the noise sources and wearing ear plugs, but the increasing density of noise and sound information is a growing fact of any expanding civilization.

It may be easy to intuit the impact of the "unwanted sounds" of our techno-urban cacophony, but some of its effects may be more subtle and harder to understand. In urban settings, along with the common sounds of our activity, we are bathed in the constant soundfield of our technologies. The preponderant hum of electrical power in the US is 60 Hz (cycles per second)—approximately "B flat" on a musical scale (57.02 Hz). In Europe, it is 50 Hz or "A flat" (50.8 Hz). Other appliances and equipment set up other tones—microwave ovens, refrigerators, pumps, computer hard drives, fans, lawnmowers etc.

Musically speaking, the incidental sounds in our environment set up chords and musical intervals. We have known for centuries that chords and intervals affect our senses and emotions—it could be said that the study of Western harmony is a study of how musical intervals affect our ideas, our emotions and our spirit. Plato wrote about this in his *Republic*, "…when the modes of music change, fundamental laws of culture change."[24] The power that intervals and musical scales have on our emotions drove the discussion of music in the early Christian church—what intervals and musical modes were permissible in prayer and worship, and how much "pleasure" could be included in the canon. The discussion continues to this day in musical circles. One of the informative issues that the late composer John Cage[25] brought us with his "composed" pieces of silence is that when we behold the soundscape and focus on all of the elements in it, significant meanings, sensations and feelings arise. Cage gave us permission to observe how we are affected by sounds that were not "intentionally" generated by musicians. It is not surprising that the discussions of his original performances were often peppered with comments about the sounds of motorcycles, jets, traffic, air conditioners and other sonic artifacts of our techno-logical culture. Through Cage's framing of our mechanistic, technological sound-scape, we find ourselves serenaded by an orchestra of soulless musicians set to play the dominant tunes of our times.

There have always been the incidental sounds of human culture—bells and chimes, the spinning of wheels, the lowing of cattle, and the pounding of grain, soil, and iron. The sounds of preindustrial civilizations were driven by human needs and the providence of nature. The industrial revolution, in changing the idea of produc-tion and need, marked the beginning of a time when the sounds of civilization were set to happen on their own, without the constant supervision of people. We have since taken these noises for granted, resulting in our being subjected to the affects of a background din that we don't entirely control.

We might want to attribute the "noise problem" to our modern industrial age, and it is true that these modern times have brought a cacophony of new and strange sounds into our environment. But our lives may not actually be noisier—at least in urban settings—than they were in the past. Lest one think that times were once quieter in the preindustrial cities, the din of metal cart wheels, horse hooves clomp-ing over cobblestone roads, drayage practices, ware monger's whistles and shouts, neighborhood iron smithing, and the yelling of the urban inhabitants competing with all of this noise—was incredibly loud in the metropolitan areas of previous centuries. According to writer Stephen Coleridge, the noise in mid nineteenth cen-tury London was "tremendous"; it was perhaps worse than might be encountered in all but the most hostile of contemporary noise environments:

All the main streets were paved with stone blocks, and as there were no India rubber tyres, the noise was deafening. In the middle of Regent's Park or Hyde Park, one heard the roar of traffic all round in a ring of tremendous sound; and in any shop in Oxford Street, if the door was opened no one could make himself heard till it was shut again.[26]

A century earlier in Paris the traffic noise was so pervasive, reverberating off all of the tight building faces in the narrow streets that pedestrians couldn't easily locate the source of dangerous horse drawn carts and wagons. Pedestrian traffic fatalities became such a problem that more noise had to be added to the din just to provide for the safety of the pedestrians:

Toward the end of the eighteenth century, Parisian authorities ordered that all horses be provided with sleigh bells so that pedestrians might be warned and be given a sporting chance to save their lives[27]

Even in Julian Rome, noise on the streets proved unhealthful to metropolitan inhabitants. In an attempt to alleviate urban traffic, Julius Caesar decreed in his Senatus Consultum of 44 B.C.E.:

> Henceforward, no wheeled vehicles whatsoever will be allowed within the precincts of the city from sunrise until the hour before dusk…Those which shall have entered during the night, and are still within the city at dawn, must halt and stand empty until the appointed hour.

To which Roman satirist Juvenal quipped:

> …it is absolutely impossible to sleep anywhere in the city. The perpetual traffic and wagons in the surrounding streets… is sufficient to wake the dead.[28]

Earlier still, the noise of humanity became so unbearable that the Gods took matters into their own hands:

> In those days the world teemed, the people multiplied, the land bellowed like a wild bull, and the great God was aroused by the clamor. Enlil heard the clamor and he said to the Gods in council, "The uproar of mankind is intolerable and sleep is no longer possible by reason of the babble." So the gods in their hearts were moved to let loose the deluge. (From "The Epic of Gilgamesh" Sumerian poem, c. 3000 B.C.E.)[29]

If there is a quantitative difference in the noise of our contemporary soundscapes, it is in the overall global density of noise. With the burgeoning population of noise-making inhabitants on this planet, there are fewer landscapes where silence and quietude prevail. The explosive compression of steam engines—once the clarion in the horizon announcing centers of industry and progress, has given way to the throttle and hum of internal combustion engines and electric motors, which we now hear everywhere, and almost incessantly. I live in a rural environment and scarcely 5 minutes passes without my hearing the sound of an airplane or a truck somewhere in my soundscape. Two small planes and a passenger jet have passed over me just since I started crafting this last paragraph. I can safely speculate that this same environmental condition holds true for almost anywhere this book is being read.

Take a moment sometime, perhaps 5 minutes, with the "hear where you are" experiment mentioned in the beginning of this chapter and count the number of occurrences that mechanized sounds and noise enter into your environment. Listen specifically for car horns, brakes squealing, computers whirring, florescent lights humming, jets and planes, air conditioners, telephones and motorcycles. Try to imagine how far away you would have to be to not hear these things. As it happens, you would have to be pretty far away—or in an environment specifically designed for the isolation of sound. Natural environment sound recordist Gordon Hempton commented that even in the remote jungles of South America or the deserts of the North American West, scarcely 10 minutes passes without hearing the sound of a chainsaw, a domesticated rooster, a barking dog, a radio, or an airplane; all artifacts of our expanding population around the globe.

A number of years ago I was planning to move away from metropolitan Los Angeles and into a more rural setting. Initially I thought that I would still like to remain close enough to the city to continue doing business with my L.A. clients, as well as take advantage of the abundance of culture this very large megalopolis offered. There are some small towns in Ventura County, just north and west of L.A. County which met my environmental criteria and were within acceptable driving distance of the city. On one of my house hunting trips I had an experience that drove my decision to leave Southern California altogether. It occurred while I was standing on the beach in Ventura facing out toward the ocean. From my right, up the coast and north-west I could hear the gentle waves on the shore and the light wind across the sands off to quite a distance; from my left and south-east toward Los Angeles I could hear a deep, profound, quiet but earth-gripping rumble. I realized that while I was more that 30 miles from the dense housing and freeways of L.A., the thunder of the city was still molding my horizon.

What is the effect of this deep, subtle rumble across our horizons—or any gratuitous environmental noise for that matter? When we are submerged in noise we lose our sense of time, place and continuity. Because it takes time for sound to unfold we subconsciously mark time in the course of perceiving sound. You might say that sound substantiates our perception of time—from footsteps and heartbeats to the pace and spaces between words as they fall into conversation; from pulsing sounds to continuous drones. These sounds all give us a sense of time, pace, and progress. So when all of the spaces between desirable sounds are filled "wall to wall" with a slurry of noise, it takes the "breath" out of our auditory experience. This probably accounts for the fact that people who work in or are otherwise subjected to noisy environments have a poor sense of time. The noisier the environment, the more extreme the lost sense of time.[30]

In the rural or wilderness settings of preindustrial times, sounds and noises ebbed and flowed with the turning of the days and nights, and bloomed and subsided over the course of the seasons. But those bygone natural frontiers have largely been invaded and consumed by our industry, our population, our agriculture, and our need for elbow room. And nowhere is it more apparent than in our soundscapes. It would be hard to determine the full impact of this noise on our psyche. Our cultural perception of time loss—of not having enough time to get anything done, and losing our community connections may be the most stressing artifacts of the saturation of our soundscapes with ever more arbitrary noise. With our personal liberty of engaging in communication at any time we wish—through cell phones, the Internet, television, radio, answering machines and such, time in the age of communication is increasingly unmediated by social, community, and geographical constructs. Family meal times, sleep and waking cycles, weather, geography, and other factors that previously limited our access to other people's time no longer impede our ability to express our will whenever we want. But this unmediated liberty is filling up with noise. And while we may be a bit less fearful, knowing that some form of humanity is always within earshot, we may also be losing our larger sense of wonderment, our feeling of comfort and humility that nature is irrefutably larger than we are. The invincible momentum which guides us all to our destiny is no longer the mystery of creation, but rather the chaos of our accumulated individual noise.

Warning, Alert, and Alarm

Sounds and noises have qualities that permeate our subconscious which affect our emotional state of being. This is most clearly represented in how alarming sounds bring us immediately to attention. From the old French *alarme* and Italian *all' arme*—or "call to arms," an alarm will cause us to act; we are not called upon to interpret or speculate. Because sound reaches so rapidly into our subconscious compelling us to move, it is used intentionally to divert our attentions away from linear thought and jar us into action.

Because the value of a military is measured in the currency of alarms, it is no surprise that the US military has explored alarms in depth. Their investigation includes many studies comparing the relative efficacy of visual and auditory alarms. What has been universally found is that auditory alarms will cause a person to act faster than visual alarms,[31] even when the subject is prepared for a visual alarm and an auditory alarm is substituted without forewarning.

Researchers have also found that in an auditory field of beeping, blips, and buzzing, a human voice alarm will take top billing—with subjects paying more immediate attention to the sound of a woman's voice than the sound of a man's voice.[32] This is due to the fact that the typical bandwidth of a woman's voice is centered in the most sensitive range of our hearing (children's voices also fall in this

Fig. 1.3 F-17 cockpit (courtesy of Lockheed-Martin)

range). If we approach the human organism from strictly a biological sense, this heightened sensitivity to the voices of women and children may have to do with their limited mobility in the years of child bearing and rearing, consequently defining the center of the village. While the men silently patrol the perimeter of the village, a woman or child crying out in fear would indicate a threat within the perimeter and thus a need for immediate action.

During the late 1970s when much of the research on voice annunciation was filtering into the public market, a few automobile manufacturers included a voice feature in their cars. In its most common form it was a mature but comforting woman's voice gently reminding the passengers to fasten their seat belts. In deluxe models the lexicon included reminders of the status of headlights, doors ajar and keys left in the ignition. The feature didn't continue for long. I suspect that having to meet with the consistently pleasant, un-modulated "voice personality" of a vehicle wore thin rather fast. It may have been that all car owners didn't want to relate to their vehicle as a white, mid-30s, intelligent, agreeable woman (a voice model probably derived from the market-typal sound of "Mom"). Folks are temporarily amused, but didn't seem to welcome conversational relationships with their cars. Even in these days, when the technology is readily available for voice recognition and conversant computers, the only places we commonly find conversational equipment is on the telephone, where the voice is the only "handle," and in cockpits of jet fighters and commercial airplanes—where the dimensions of emergencies are potentially huge.

Sound is so effective in getting us to move due to the way we perceive and process auditory cues. Sight works in straight lines—we need to direct our vision to something in order to perceive it. On the other hand, hearing is hemispherical; we hear sounds from all around, not just where we cast our ears. Because we continuously gather sound from all sectors, we are very sensitive to minute changes in our soundscape. We are literally wired to be hypersensitive to acoustic changes. Our ears keep vigil on our surroundings, immediately perking up to any new stimulus. Once our brain identifies and accepts an audio event, it backs off its processing power and awaits the next new event.

Along with our hypersensitivity to change, we are also made more auditorily alert by rich harmonic content. With very few exceptions, all sounds contain more than just one simple frequency or pitch; most sounds contain a fundamental sound—a thump, clang, or a pitch, simultaneously mixed in with other characteristic sounds. The mix of these sounds create the "timbre" or quality of a sound event and help us distinguish the difference between someone talking and the same person yelling; or the difference between a mountain lion screaming and a baby crying. When something is far away, the higher frequencies of the sound are attenuated by the environment, so an African lion roaring in the distance sounds more like a rumble than a roar. As the sound gets closer, the higher frequencies become more apparent. A lion roaring on the verandah will have a rich mix of high and low frequencies (and a thick concentration of terror) while the distant low roar might only make us anxious. In the verandah case we are tuned to identify the rich harmonic mix of the roar with proximity—and threat.

A similar sound timbre characteristic exists with all vocal expressions—from monkeys to birds, from lizards to humans; as their expression gets more urgent, their voice gets louder and more forced. When a voice is forced it becomes more harmonically rich and contains more high frequencies, so we associate harmonic complexity with urgency and become more alert with the increase of harmonic content. This "harmonic content motivator" is easy to grasp from the perspective of hearing voices. It is also used when designing alarming sounds. A residential fire alarm is designed with a rich harmonic profile to cut through our deepest sleep and get us to jump and run. The fire alarm needs to immediately command our complete attention, unlike the pleasant repeating chime used in automotive interior alerts which would be ineffective as a residential fire alarm, even at high volume.

The "harmonic content motivator" of alarms also occurs in our built environments. We will tend to be more alert—at least initially, in a room built from hard, sound reflective surfaces with a concentration of reflected high frequencies, and less alert in a cozy sound absorbing room weighted with harmonically dampened low frequencies. Spending time in a highly sound reflective environment can become fatiguing; we tire of subconsciously identifying each new sound event as if it were important. Conversely, a sound-absorbing environment can end up putting us to sleep. To get a sense of this, consider the relative feelings of "alert" or "cozy" you get in a kitchen to those you get in a bedroom; in the clatter of a restaurant to the hush of a movie theater; in a bowling alley or in a card room; from sitting by a waterfall or walking in a forest glen.

Our feelings of alertness are modulated by our biological need to pay attention to alarm cues imbedded in the acoustical profile of our surroundings. These cues of harmonic content, loudness, and uniqueness of sound inform us, and are mediated by our circumstances. In the pre-technical environment of a hut, cave or longhouse, the acoustical setting is supported by the intention of the space, and alarm cues are easily reconciled with the circumstances. But as our built environments become more complex (and designed more for how they look than by how they sound), we risk building perceptual conflicts into our surroundings that may have the acoustical profiles of "alarm" which don't support the true intentions of the space. By way of example, if the design intention of a space is "focus and concentration"—such as a library or study, and its acoustical profile is bright and sound reflective, we will be distracted by the surrounding active noise field. The "over alertness" induced by this edgy, reverberant environment is anxiety producing. This anxiety is partially due to our not being able to hear new information buried in the surrounding noise—new information that may bring us an unpleasant surprise. Subconsciously we are constantly evaluating alarm states outside of the perimeter of our focused interest.

Alarm signals are designed to bring us outside of our conscious field of interest by using incongruous noises. Bells, whistles, claxons, and sirens are all used because they are louder, and cut through the sonic environments they are designed to protect. But there are conditions whereby unexpected silence—and the breaking of a comfortable soundscape—is just as alarming as a loud and nasty noise. This "silence alert" is deep, and springs from the archetypal fear all creatures have of being stalked.

When creatures suspect something threatening in their surroundings, they will immediately quiet down in order to focus in on the sounds they aren't making. We subconsciously rely on this silence alert to help us monitor our unseen surroundings. Perhaps the most telling example of this involves our comfort in the sound of the night chorus of crickets. While out walking at night, the singing of the summer crickets is assuring (whether or not we consciously make the association that crickets will chirrup only when they themselves are comfortable with their surroundings). When the crickets suddenly stop singing, we are alerted; anxiety replaces comfort, hair rises on the back of our necks—we get the creeps. A cricket's "silence reaction" to something new coming into their environment is so predictable that they were once used for night sentry duty in Japanese homes, where having a cricket as a housemate was also considered good luck.

> In the Orient crickets are often kept as pets, replacing watchdogs, for even while fiddling a steady summer symphony a cricket is listening, his wide acoustic range sensitive to even subsonic ground vibrations. The musicians will cease in unison during a harmonic rendering if someone unknown enters a household. The sudden silence awakens the sleeping family.[33]

While our conscious response to noise or silence is largely dependent on our familiarity with the setting, our subconscious responses—our emotions and feelings of comfort and safety—are influenced by the other textures of the environment. We are affected by such factors as how bright or dull our surrounding sounds, how much detail we can glean from the soundscape and how well we can assess distant sounds. If we are able to survey our soundscape much in the manner that a scout surveys the distant horizon, we can assure ourselves that nothing pernicious will come lunging out of it. This is particularly important at the onset of night.

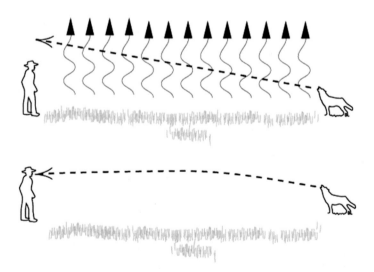

Fig. 1.4 Defraction. Rising hot air causes sound to deflect upwards, sinking cool air deflects sound downward

As night closes in on us, so does our world of perception. Fear, mystery and intrigue prevail; we don't know what is just outside the perimeter of our perception. The darkness limits our ability to visually perceive distance, and when the night first falls our auditory perception of distance is limited as well. During the day, the sun heats up the earth's surface. Warm air rising up from the earth refracts sound up into the atmosphere directing distant sounds up above our heads. This rising warm air limits our ability to hear long distances—a condition that remains long after the sun has set. We sense this limitation subconsciously, and accentuated by the darkness of night, it activates our feeling that there are things occurring within our physical realm but outside of our perceptual reach. At some point during the night, the earth surface cools causing the cool air to settle and distant sounds to adhere to the earth surface, so as the dawn breaks to shed new light on our horizons, we can also hear greater distances—capturing more of our surrounding world. Of the many perceptive cues that signal the safety of a new day, our ability to better hear our surroundings renews and bolsters our confidence; we can somehow contain all that is in the perimeter of our hearing, and if we can hear a broader horizon, it diminishes the likelihood of our experiencing unpleasant surprises coming out of it.

Sound, Comfort, and Belonging

Ever alert to signals of danger, we also seek acoustic assurances of safety—represented in the sounds of shelter and of laughter, of cooking and of song; the pattering rain on the roof and a crackling fire in the hearth. These comforting sounds—almost as indelible to our species as the sounds of heartbeats or breathing, give us messages of security, grounding, abundance and growth.

Sound's fluid quality saturates and surrounds us and is by this nature inclusive. We hear things and are included with those things in a common soundfield—as both participants and inhabitants of our soundscapes.

Sounds can affect our sense of well-being, comfort, and safety—even in our "pre-cognitive" infancy. Dr. Lee Salk studied infants in maternity wards exposed to three different sounds; the sound of lullabies, the sound of a ticking metronome at heart rate, and the sound of heartbeats. In the three groups of babies, it was shown that the heartbeat group clearly fared better in falling asleep (taking half the amount of time), the times when at least one infant was crying (38 % as opposed to 60 %), and weight gain.[34] This is not to say that the other groups fared particularly badly—though there was no group mentioned in the studies who only heard the humming of the fluorescent lights and the sound of their ward mates crying—the soundscape which might be readily found in modern maternity wards.

Newborn babies, limited as they are to seeing what is directly in front of them, first explore the world through their ears. While they may not understand all of the conversations we present to them, they have their auditory feelers out, sizing up their new world. By the time a child is born they have already had a 3–4 month auditory preview of the place they will inhabit, so it should be no surprise that they respond to familiar voices and sounds. When Dr. Brian Satt was developing his

"Womb Song"[35] program mentioned earlier in this chapter, his study included intra-uterine recordings of the "outside world" just prior to the mother delivering her baby. The recordings have a surprisingly clear audio fidelity; conversations, squeaky doors, nurse calls, and singing all yielded with the clarity of a radio to the accompaniment of heartbeats, blood flow, and digestive gurgling—quite a swinging scene.

As we grow into—and become part of a family, and then part of a community, the sound of other people's voices comforts us with the feeling that we are not alone—that we belong to a diverse tapestry of other lives. In the efficiency of our current society, we have thwarted this diverse connectivity to some degree by segregating our institutions—such as schools and healthcare facilities—into age or process classes. The commodified efficiency of infant and childcare facilities, as well as elderly nursing homes hinge on isolating those who need care and oversight from the more "productive" members of society. From an efficiency standpoint this makes some sense, but knowing how the sounds of belonging promote health, hope, and well being, this strategy presents some drawbacks. But there are some healthcare institutions—hospitals, eldercare facilities and such that do make a conscious effort to include the sounds of community into their convalescent wards.

When preschoolers are brought into convalescent homes, the sound of children laughing and playing lifts the spirits of the elders, just as the sound of the elders speaking is assuring to the children. Intergenerational health care facilities are beginning to make appearances even in the more "rigid" context of corporate health, promoting an integration of all ages of people within their care. Brought together, patients feel like they are in it together, not isolated "freaks"—aberrant invalids outside the society of the "healthy." If we can reestablish community ties which have been eroding in the past few decades, we may find the task of caring for our families much easier.

> Children seem to be able to reach through the infirmities of age in ways that adults cannot. They tend to have more physical contact with elders; their playful activity, even if watched from afar, makes seniors feel like participants in what's going on around them. At St. Francis [Gardens, Albuquerque NM], children have helped renew despondent residents' interest in eating. And ["Generations Together" executive director Sally] Neuman found that a group of normally unresponsive Alzheimer's patients smile, moved around, or tapped a beat during music activities when children were present. A group listening to the same music without the children was unresponsive.[36]

A little excavation into the dictionaries and etymologies reveals the Anglo Saxon word for "wholeness" imbedded in the word "health." Surely this includes patients not being segregated from the social organism to which they belong. Digging a bit deeper reveals the word "health" is also inextricably tied in with the adjective "sound," where "heal" is "to make sound, well, or healthy again and restore to health." "To grow sound; to return to a sound state," and where "sound" is "whole, unimpaired, unhurt, not weak, diseased or damaged..." Wholesome.

It is sound that bonds us to our community and keeps us whole—and not just by way of speech and conversation; we hear the movement and actions of those around us—distant voices, footsteps, cars passing, the sound of running water in our pipes—all indicating the presence and proximity of others. In a family dwelling or in a community of familiar and friendly people these noises are comforting, assuring us that health and safety are within reach.

It is on account of this "strength of community bonding" through sound that I was deeply saddened by a comment my grandfather made in the closing years of his life. He said that while the inevitable age deterioration of his body was hard, he was most challenged by the gradual, and eventually the profound loss of his hearing. "When us old folks find it hard to walk, or hard to see, people reach out to help. They open doors and carry our loads. But when we lose our hearing, people can't speak to us, they think we're stupid." Adding this insult to the erosion of his auditory connection with his community presented a sad specter to me—highlighting the importance of maintaining the health of our own hearing.

We will all experience deterioration of our hearing as we age. In our current industrial culture, over 20 % of folks over 65 years old have a significant hearing loss—a loss that presents obstacles to clear and easy communication. It would be hard to determine if this inevitable hearing loss is better or worse than in preindustrial cultures and times—given the differences in medicine and life expectancy of various peoples throughout history. What is clear is that with the advent of electrically amplified music and sound, we are accelerating our hearing loss at an alarming rate. A rather tragic indication of this was brought home in a study of hearing loss done by English auditory researcher Edward Evans:

> Evans has for decades used his students at the University of Keele as his "normal" group in research on noise-induced hearing damage. "About seven years ago," Evans says, "we began noticing some of them had *worse* hearing than the old fogies in the department." To see whether the changes were caused by loud music, Evans tested young people who throughout their lives had listened to varying amounts of music…When Evans checked the students "filtering" ability—how well they heard a particular sound against a background of competing noise—he found that those who'd listened to a lot more music showed a 10 to 20 percent deterioration, compared to the students with a quieter life. That isn't enough for the students to notice, but it worries Evans. In modern, noisy cultures, middle aged people often have trouble distinguishing speech in situations like cocktail parties, where hubbub competes with conversation. Evans fears the students will develop such problems decades earlier.[37]

It is probable that some of this hearing loss is simply caused by living in noise-saturated environments. This common urban aggravation is likely accentuated by ever more frequent contacts with amplified media—concerts, radio and movies, and sound amplification through the ubiquitous "personal listening devices"—miniature headphones which are a particularly egregious source of hearing loss. These devices couple sounds directly to our ear's tympanic membrane. The physical cues that warn our bodies of harmful sound levels are not present when we use headphones. To get the visceral sensation of loudness with headphones alone, the volume really needs to be "cranked up."

Perhaps we won't attend to this problem until we really feel its impact on our society. Outside of direct testing, gradual hearing loss is hard to detect except by behavior—and the artifacts of this behavior are just as gradual and just as hard to pinpoint as the loss. A person will increasingly misunderstand conversations; they may over-modulate the volume of their voice, speaking too loudly or too quietly for a given situation; they may listen to amplified sound louder than practical—turning the television or radio volume up to levels that annoy others. Perhaps the most sinister artifact of this gradual hearing loss is that a person—not feeling in auditory contact with their community or family—will withdraw, isolated and alone, into the internal soundscapes of their own imagination.

Hearing, more than any other perception, is the perception of inclusion. When we lose it, it erodes our sense of participation. We are sensually woven into our surroundings and our community in the fabric of sound. We know this when we speak, we feel this when we listen; but even when we are not consciously attending to sound—while we sleep and dream, we are still included in its compass.

Media and the Sounds of Our Times

It has been said that by the late twentieth century, we humans processed more "new" information in a week than our seventeenth century forebears processed in a whole adult lifetime. The simple act of driving down a freeway while listening to the radio talk show forces us to be many places at once; our imagination is on our destiny, our concerns are on the talk show topic, our linear thoughts are mapping our progress, and our heart may not be in any of it—all the while blasting along at 65 mph. The processing density of our perceptions sets the tenor of our media driven society. We are learning to process information in multiple layers. We can watch a TV newscaster speaking from a "news desk" set with a cut-in of a disaster on-screen behind her and a mix of "live remote" sounds, subtitled with statistics and locations. Video games have control panels with ever increasing analog information; a churning of "music" soundtrack, coincident sound effects, voiced responses, as well as the visual content of the game. Even the Internet—a library upon which we are becoming more dependent—targets our ears, our eyes, our minds, and our wallets all at the same time.

Each one of these layers is crafted toward one objective—to capture our full attention so that the imbedded messages can be conveyed without external distraction. Increasingly, layers of sound are filling in the cracks where our unattended imaginations might wander. Film and television productions have always had three basic fields of sound: dialog, sound effects, and music. In the last two decades of the twentieth century, the roles of these three fields have matured and transformed the media experience. Dialog is still central to most productions, and film music continues to be written both for dramatic support as well as for its "after-market" value, but the world of sound effects has truly taken a leading role in mediating the emotions of viewers.

In the early days of radio, sound-effects artists would support radio drama with a battery of live sounds—using water whistles to suggest birds, squeaking balloons to evoke creaking door hinges, and coconut shells to suggest galloping horses. While these early sounds were clearly "not real," they were persuasive to the imagination. Given the innate fertility of our imagination, these early radio sound effects were adequate for the task of engaging the listener, propelling them into appropriate emotional spaces. But with the development and addictive popularity of moving pictures, sound effects needed to be ever more persuasive—due in part to the sense of realism that the visual medium spawned. With the limitations of the flat cinema screen and "compressed" sound amplification,[38] both the visual and audio tracks have to be carefully crafted and framed so that media sound isn't heard as "sound," rather it is heard as if it is just part of the scenery.

Most people are surprised to learn that in most finished Hollywood productions maybe only 5 % of the sound heard in the theater occurred on the shooting set. The remaining 95 % is recorded and assembled much later in specialized studios, including the dialog or spoken parts. This is due to the dramatic need to enhance or create sound environments which, though unrealistic, are more persuasive than what can be procured by recording the actual sound on a live movie set. In "post production" the sounds of footsteps, wind and rain, automobile engines and spaceships are carefully layered and constructed out of the recorded sounds of peas falling, gurgling aquarium pumps, "plicking" paper cups, dropping watermelons, charcoal burning on plates of Pyrex and high tension wires being bowed. The film industry name for the creation of movie sound effects is "foley," named after Jack Foley, one of the first artists in this field who was particularly resourceful in his understanding of how we perceive recorded sound. Many improbable sound combinations are assembled and recorded by "foley artists" who may bring a collection of objects, gadgets, and things to a sound stage (suggestive of a psychotic's picnic basket).

Academy award winning sound designer Randy Thom states that sound "… sneaks into the side door, so to speak, and goes straight to the heart of the listener, avoiding the kind of analysis we tend to apply to visual images…"[39] By going straight to the heart of the listener, a film maker can lead the audience away from a conscious engagement with the mechanics of a film and more easily transport them into the suspension of disbelief necessary to pull off a persuasive experience.

Given that much of the media's message is targeted to the suspension of our disbelief, constantly subjecting ourselves to this suspension can modify our true beliefs. While engaged in media, we may be subconsciously compelled to feel pride, adventure, anger, or fear and not know exactly why. If we want to get a feeling for the impact that media has on our life experience (and our resulting beliefs), we need to consciously isolate and focus on each part of it. One way we can do this is by consciously removing it from our experience—stop listening or watching. A proactive way to explore media impact is to systematically modify or disassemble the elements of an assembled presentation. We might listen to a television game show without the picture or watch a sitcom without the sound, play a video game while listening to the broadcast news as a soundtrack, or watch a soap-opera while listening to Ornette Coleman. In this disassembled or reconstructed state the associations take on new meaning, the carefully designed significance falls apart and the produced myth is revealed.

The intrinsic value of media production became clear to me in the early 1980s, while working as a sound designer for the soundtrack of the movie *Koyaanisqatsi*,[40] a movie that focused on reframing the assumptions that make our technological society possible. In the course of our work I had an opportunity to isolate specific sound elements imbedded in popular mass media. We recorded a large array of sound sources from various media resources—radio talk programs, television game shows, sports announcers, broadcast news, "Christian" television, advertising, sitcoms, science programs—the whole bit. From these we extracted specific "sound bites" and assembled them into thematic reels. These sound bite reels were assembled under specific categories. We had a reel called "Money" which included only the quotations of prices pulled from stock quotes, real estate and automobile advertising, game show winnings, and product offerings. Another reel called "Goodies" consisted of recitations and exclamations of consumer items lifted from similar sources. Other reels included "Food," "Time," "Sex"—covering the basic ingredients of contemporary culture through their sonic exclamations in media.

While the reels contained snips of phrases and words, each one of which carried meaning, what was remarkable was that the overall sound of each of these reels had their own impact, accentuated by the fact that the elements were excised from their context and presented as a whole. The almost sacred gravity with which stock quotes resonated; the forced levity announcing consumer items; and the droll, lugubrious recitations of times and dates spoke volumes about how our culture perceives these categories.

Isolating the categorical phrases accentuated their dramatic content, so when we played our assembled reels to the production staff everyone got a good laugh, but by far the highest emotional impact was derived from a reel we called "Death and Disaster." This reel consisted of key phrases excised from news and dramatic productions dealing with catastrophe. It was a 10 minutes continuous string which played—"… A cloud of poisonous fumes—100 people died—the bullet hit the body, hit the lower body, hit the body—shot the Pope—police action today—rioters attacked—injury accident on northbound—people still buried—small plane crashed—"how could it have happened!?"—lost hikers—beaten by the angry mob—bleeding still—broken limbs—died today—body count—terrorist bombing—let the war begin…" The piece dropped a dark pall over our group. The producer didn't even want to use this reel even though the purpose of the piece was to illustrate the impact that these "subtle" messages have on our collective psyche.

While we are continuously being pummeled by these phrases, we believe that we can shut them out, all the more easily to have them "sneak into the side door, so to speak, and go straight to the heart of the listener, avoiding the kind of analysis we tend to apply to visual images…"

As the fabric of our experience stretches to fit the matrix of our technological society, the sounds that we subject ourselves to by way of commercial media are becoming an unacknowledged subtext of our culture. With the background din of produced media defining the perimeters of our lives, what we assume resides at our perimeter may actually be more a fabrication produced by others than a direct consequence of our own encounters. Television and radio sounds are increasingly becoming a fixture in our soundscapes: family rooms are dominated not by the hearth, but by the big-screen TV; public commons such as bars, restaurants, waiting areas in airports and hospitals all include the ubiquitous screen with its attendant murmur. It is not uncommon for people to turn their television or radio on—not for information or entertainment, but for companionship. There may be something comforting about hearing the sounds of familiar voices in the background, even if what they are saying makes little sense to our true surroundings. An unfortunate consequence of this is that the dramatic events and discussions presented in broadcast media—tailored to incite, excite, stupefy, and sell—have little bearing on the texture of experience in a real community or family setting. We may end up fearing the actual experience of community because we subconsciously believe that it is accurately depicted in our media as fearful. Our experience of our community, culture and civilization is increasingly being sculpted by a background soundscape that we are not really listening to.

As cacophonous and confusing as they may be, we cannot eliminate the sonic artifacts of our culture; they are as much of an expression of where we are today as the jangle and clip-clop of draft horses, the unfurling of sails and the firing of the noon-day gun were sonic expressions of bygone eras. The difference however, is that sounds in these earlier times occurred as a consequence of specific events, inducing emotional responses appropriate to the setting. Media produced sounds on the other hand are already bundled with intention; they are a consequence of conscious choices by someone else at some other time, with the emotional content an imbedded property of the sound.[41] With this in mind, we might try to consciously hear our soundscape—really listen to it. By considering the sounds of our time and of our culture, we can exert some control over it, allowing us to feel less detached and more aware of where we dwell.

For most Americans there are only two places in the world, where they live and their T.V. set. (Don DeLillo in *"White Noise"*)[42]

Sonic Imprinting: Acoustic Communities and emotional responses to specific sounds

There are certain sounds that will consistently have a deep emotional impact on most members of our species. The crying or cooing of a baby, the whimpering of any infant animal, the plink and splatter of a spring rain, the gentle trickling water and the warm crackling of a fireplace are just a few of the sounds which stimulate predictable emotional responses in most people. The predictability of our responses probably hinge on the fact that these sounds all cue us into sensations of abundance, procreation, and comfort. As soundmarks in our landscape of security and emotional safety, we feel safe when we are within the compass of these sounds. Conversely, when we are within the acoustical field of deep rumbling and explosions (sounds of things larger than ourselves), nasty screeching, sirens or loud reckless rattling (sounds of things out of our control), we are likely to feel anxious or threatened. The predictable human responses to any of these sounds indicate something about our common experience of the safety or vulnerability.

Along with these "universal" sound cues, there are also sounds particular to regional experiences of a place that, while not shared by all of humanity, do bond groups of people in their "acoustic communities." The lowing of cattle and the murmur of hoof beats on an open plain have a place in the emotional terrain of cowboys and cattle herders, though it would probably take a city dweller for an anxious turn if they suddenly heard these animal sounds out in their streets. The city dweller would not have reconciled these sounds to their surroundings so they would be very inconsistent with their idea of abundance and safety, and incongruous to their experience of their regional soundmarks. On the other hand, the city dweller would likely feel more comforted by the purr of a distant freeway or the clatter and bang of the trash being picked by the morning trash collectors.

The lexicon of our life experience involves collecting perceptual events and situating them on our map of understanding and response. Sounds have an informative role in this process because once we have experienced something its sound can become an encapsulated cue to the recollection of that experience. The human predilection to formulate words is part of this aspect of sound perception; we create iconic sounds that cue us in to recollections of experience. While we have assembled vast vocabularies that represent our common experiences with words, there are equally vast sets of sounds experienced by groups, clans, families, and individuals that never come into the currency of symbolic representation through words; they remain sound cues that induce responses unique to the specific listeners.

These sounds can be as distinct as the peal of particular set of church bells or as ethereal as the way the sounds of a village fill the valley in which it lays. The common soundmarks of a group do more than establish a prosaic bond to their sense of place; it also supports a sense of continuity that the group may require for their survival. In a Brittany fishing village, for example, a freeway that was planned to skirt the town was rerouted when it was determined that the noise of the freeway would drown out the sounds of the changing winds that the fishermen required to assure their safety at sea.[43] While these fishermen may have not articulated their need for their soundmarks prior to the freeway question, it was a sure thing that the winds would rouse them from their sleep, or put them in a mood when the sound of their changing triggered a need for some particular action.

Unfortunately in many situations, this sensitivity to the importance of sound is not a driving factor in the decisions of economy and progress. In the words of a Penan elder from Borneo:

> We yearn for the sounds of the forest. We have always heard these sounds. And now it is harder for us because we hear the sounds of the bulldozers. And that is what we always talk about, we women, when we get together. How will we live, how will we thrive now that we have all of the problems? To us the sound of bulldozers is death. We weep when we see the forest that has been destroyed...[44]

When events with strong emotional density occur, the auditory signatures that accompany the events imprint into our psyche. For the Penan women, the sound of bulldozers has become the sound of death. Even if these women survive to live in the city, the sound of bulldozers will have none-the-less imprinted deeply into their hearts. Similar statements can be made about the searing roar of low altitude fighter jets for Afghani and Eritrean villagers and the sound of attack helicopters to Vietnamese and Colombian villagers. Long after the threats of conflict have left these people, the imprint of the sounds will remain to shake them from their dreams.

Fortunately, our auditory imprints are not exclusively littered with the detritus of trauma and catastrophe. Most of our imprinted sounds—sounds that we largely take for granted—are sounds that confirm our daily experience, and have deep motivating potential. This is clearly represented in the auditory bonds between parents and children. Almost all parents have an extraordinarily acute focus on the sounds of their young. Nursing mothers will spontaneously lactate on hearing their baby's whimper of hunger, and fathers will perform superhuman feats when hearing their lost child cry out of a crowd of thousands. The reflexive holds true as well; children often hear their parent's voice even before the parent has formulated something to say. The preparatory clearing of a throat or smacking of lips is enough to set the child into a pout-or-flight response.

While humans may have a more complex imprinted lexicon than most creatures, the phenomenon is not species-specific. Cats will come running at the sound of an electric can opener; and upon hearing the distant sound of a specific car, dogs will head toward the garden gate to await the arrival of their human. Imprinted sounds, even out of context will trigger emotional responses in creatures:

> The Liedenpohja dairy herd recognizes many signs of spring; different sounds and smells, for example, like the smell of the new green leaves. The cattle become very excited when the big doors are opened, letting them out into the summer pastures. In one cowshed the sound of summer was once heard in the middle of winter; there was a cuckoo singing in the radio. Kerttu tells that the mouths of all the cows dropped open and chewing stopped, while they listened, astonished: "Is this summer?"...[45]

Due to the subconscious nature of our sound perception, sonic imprinting occurs at the intersection of our hearts and our minds; it is an important aspect of our auditory maps. We build up a lexicon of sounds which help us recognize safe and unsafe situations, good and bad outcomes, fearful and delightful anticipation. The sound of our lover's car pulling into the driveway; the sound of our mail slot slapping shut; the sounds of water running in the kitchen; the sound of police sirens—all can evoke emotional responses of desire, expectation, comfort or dread. These sounds imprint on our psyche through the repetitions of hearing them and the associated experiences we have with them. If we heard our lover drive up in a go-cart, or heard water running in our mail box, we would not likely respond with anywhere near the depth of emotion as we do when hearing a sound from our collection of imprinted sounds.

Perhaps this may seem too obvious to mention—that the sound your mother's voice, the gurgle of your aquarium at night, the creaking step on your front porch, the school bell ringing, the noon whistle, your car cranking over, your phone ringing, your toilet flushing—are all part of your personal auditory map with the power to move you in predictable and repeatable ways. Sounds specific to your experience will not move me as deeply as they move you, though the sonic artifacts of our culture—the sound of air conditioning, telephones, computer hard drives, police surveillance helicopters overhead, and the deep rumble of a metropolis probably do bond us in a common experience of geography, time and culture. There are sound-prints of our civilization that are so deeply planted in our emotional mythology that they still move us even when they are no longer part of our common experience; the chugging of a steam engine, the clop and grind of a horse drawn cart or the baleful moan of a steam locomotive whistle—once common elements of our culture but no longer evident—will still cause us to drift into the mythical landscapes of our imagination.

There are yet other sounds so imbedded in our collective psyche that all sentient beings will have a deep emotional reaction to them: The various sounds of the elements—fires, waterfalls, rain, streams, winds and storms, rocks falling, and thunder—will always move us to our core. These archetypal sounds are obvious illustrations of how deeply affected we are by our auditory environments.

We are continuously creating and modifying our auditory maps, comparing and cataloging new artifacts, evidence, and sensations. We are always sounding into our environment, keeping subconscious tabs on our placement within it so that we can safely collect the visual, symbolic, and tactile information we need to thrive, flourish, and prosper. As we explore the realm of our auditory sensation we can perhaps build more tools to promote our success and understanding of where we are.

A late and dear friend of mine spent many years of her youth in Nepal. Her father was a western physician practicing there. She lived in a valley above which, in the surrounding mountains was a Tibetan Buddhist monastery. In the early 1970s she related some of her memories of hearing the monks ring these metal bells which sang continuously "like a glass harmonica"—rubbed wine glasses—producing rich, high pitched harmonic drones. It was almost a decade later when I heard these bells for myself.

These bells are made of some exquisite alloy. Some can be quite ancient, and when they are rubbed with a wooden wand, they do sing like wine glasses. I am told that for the Tibetans, the sound of these bells is the sound of the Soul.

The Song of Creation

2

A Tree ascended there. Oh pure transcendence!
Oh Orpheus sings! Oh tall tree in the ear!
And all things hushed. Yet even in that silence
a new beginning, beckoning, change appeared.
Creatures of stillness crowded from the bright
unbound forest, out of their lairs and nests;
and it was not from any dullness, not
from fear, that they were so quiet in themselves,
but from just listening. Bellow roar, shriek
seemed small inside their hearts. And where there had been
at most a makeshift hut to receive the music,
A shelter nailed up out of their darkest longing,
with an entryway that shuddered in the wind…
you built a temple deep inside their hearing.

(Rainer Maria Rilke, *"The Sonnets to Orpheus"* Translated by
Stephen Mitchell, 1985, Simon and Schuster, New York)

"In the beginning was the word…" and so begins one of many accounts of The
Creation wherein the fabric of the cosmos is woven by the voice of God.[1] All world
peoples from the Aborigines of Australia to the Zulu of Zimbabwe describe the
beginning of the universe through sound: The totemic beings weave the Songlines
across Australia; The Chameleons of Yemen and Madagascar sing into the primordial forest to bring the world into existence; the Quiché Mayan *Popul Vuh* tells of
the Guacamatz—the givers of light, who consult, and while they speak their deep
understanding brings forth the dawn. They speak about the forests and about the
nature of life; how the waters will flow and how crops would be sown—and these
things appear from their words. In the first hogon of the Glittering World, the Holy
People of the Diné sing the Blessing Song from which creation emerges; and for the
ancient Sumerians, the power of creation consisted primarily of the divine word.
"All the creator had to do was make his or her plans, utter the word, and pronounce
the name"[2]—a pronouncement echoed in the Koran, explaining that Allah need only

M. Stocker, *Hear Where We Are: Sound, Ecology, and Sense of Place,*
DOI 10.1007/978-1-4614-7285-8_2, © Michael Stocker 2013

to say "Be" and It shall become—dispelling any doubt one might have in divine miracles.[3]

The acoustical creation of "All That Is" is implied in the word "universe" from the Latin "one verse" (a verse being a complete turn or complete idea drawn through a line of a poem). The word *poesis* from which this poetry derived is from the Greek for "creation." Even our most advanced theories of the birth of the universe start with a Big Bang…

Sound gets stuff done. It is the perfect force for the Gods to wield, making the divine manifest in physical form. The universality of this idea is spawned by the dual nature of sound as both ethereal and visceral, and supported by our experience of the radical transformations that occur by way of sound energy. Energy borne on the wings of sound can work us on a multitude of levels; from the cognitive to the subconscious; from the emotional to the physical. It is integrally woven into our experience of the sacred, wielding profound power to affect us, while remaining something that we cannot grasp, are unable to see, and does not seem to affect the objects that surround us.

Sound needs to impinge on a living being to take effect, and the consequences can be deeply moving. Reaching into the depths of our own souls, it can instantaneously propel us to ecstasy or devastate us into dark depression. The compass of sound's influence can be vast enough to unify nations, though its effect may also be so individually focused that the very sound that you find thrilling may cause my sadness—and then just go unnoticed by someone else.

Entirely engaged in the world of sound, we have few provisions to stop it from affecting us; by the time we plug our ears to prevent an unwanted sound from reaching us, we already know the meaning it bears. The only way we can effectively prevent sound from influencing or affecting us is by making sound that is louder, more provocative, or more beautiful—creating our own acoustical world. So it is by way of sound that we too may become creators and exercise a prerogative which continuously suggests that by "singing creation," we, like the Gods can manifest the divine and create our own world.

In this chapter we will explore how and why we sound into our surroundings; affecting change and co-creating an environment that resonates with our needs—a common trait that we share with all sounding beings.

Some Songlines

Perhaps the clearest continuous connection to sound and creation is found in the life, culture, and language of the Australian aboriginals. Their land was created by the First Beings as they emerged from the earth to sing their way across the primordial plains; scraping deep canyons in search of water; heaving stone mountains across the landscape in conflict; weeping rivers, disgorging forests, and pushing up hills and dunes while making love. Every feature of the landscape has a verse in their song; the paths of these totemic beings wove all of creation together on their intersecting paths before they submerged back into the Earth. Those who inhabit the land are continuously part of the evolving map spun together by the Songlines. An expecting mother, when she first feels the quickening of her baby will take notice of where she is, conferring with the elders to determine of what Songline or "dreaming" her child will be. A child of the Rock Wallaby, the Honey Ant, or Barking Lizard "dreaming" belongs to their totemic songline in a manner more cohesive than their connection to their own birth family or tribe. As they "walkabout," they sing their songline, learn its features and legacy, and keep it alive by breathing experience and understanding into it over the course of their life. In this manner their entire world is tied together. Even those who don't speak a common language will recognize human relatives of their own dreaming across the continent through their common song.[4]

We may feel far placed from the enchanted world of the Songlines, but we may be closer than we think. The artifacts of this reality still dwell in our own language, ready to be awakened. The very word "enchanted" conveys a great deal about the transformational aspect of even a simple song. To be en-songed—submerged within song and transported outside our mundane reality.

The singing voice is the first musical instrument used to convey emotion, and it is certain that creatures were singing expressions to each other well before any symbolic vocabulary was ever derived.[5] The sound of song is still compelling enough to convey emotions not only across cultures but even across species lines.

To sing our experience and call out our perceptions is to live in an enchanted world. Being "enchanted" suggests being induced into a dreamlike relationship with our surroundings—outside of the perceptual constraints of linear time. We who have learned that songs are mostly narratives about certain events from particular perspectives may have forgotten that the original songs—like the songs of the birds—were used for courtship and community bonding, expressions of relationship and territory, or exclamations of fear, lust, anger, and arousal. Humans, with our mimetic abilities used singing to draw those within earshot into our song; seducing friends into our compass and animals into our lair. As we become more versed in singing, we use our musical voice to bring ourselves out into the world, harmonizing with our surroundings and those within it. Song engages and transforms the living beings around us, but the act of singing also transforms us from within (and is the only way we can touch our own body from inside). The gift of song is so compelling that when the Muses introduced singing to humanity, some people were so delighted that they sang continuously, forgetting to eat or drink. Socrates tells us in the *Phaedrus* that these enchanted souls became cicadas, given the gift of perpetual song from birth to death, honored by the Muses in Heaven on their demise.[6]

Human creation of song could be likened to a plant's creation of flowers. A flower is a distinct expression of the plant that dwells within the seed; it is an offering to the outside world, engaging insects in the act of pollination, and seducing humans into the act of cultivation.[7] Neither songs nor the flowers are complete expressions in themselves—they require other beings to really exist. Thus flowers form around the desires of their pollinators with a play of light, color, shape, and odor,[8] just as songs congeal around the emotional disposition of humans, attracting listeners' sympathies using harmony, rhythm, and lyrics. So while a song conveys a sense of the singer's emotion, it also plays upon the listener's sympathy, inducing their participation in that emotion.[9]

The human predilection of sounding our surroundings begins early on. The innocent, playful "tra-la-la" songs of the child—amusing themselves with their own sounds in their surroundings—are engagements with life in a manner that is woven into the organism.[10] Their cooing and burbling is a reaching out for a response from the world. They form their own sounds around the responses they receive. Sounds that evoke predictable responses are used again; new sounds are invented to test new things. Through this singing, a proto-language emerges; a language that adults may not clearly understand, but nonetheless serves the child in connecting them with their environment. Eventually this proto-language dovetails into our collaborative vocabularies so that "stone" becomes a stone, and "water" becomes wet in our mind as well as in our mouth.[11] We sound out our surroundings to affirm its existence, calling things names to place them in our experiential vocabulary. We recall the names, conferring and sharing them with others. We use these common names so we can communicate and exert some control over the things we name in a world we mutually recognize. It is by way of this naming that the song of creation becomes the sounds of engagement and the vocabulary of influence.

Naming: Taking Possession of the Created

Adam's task was to name the things of creation—exhaling the breath of life blown into him to identify "every beast of the field, and every foul of the air"—ostensibly to know them and have dominion over them.[12] This naming metaphor holds true across the breadth of human cultures, as language and sound become the medium of exchange between these experiential song-beings and their surroundings.

It must have been a delight for the first naming humans, wandering into an unsullied relationship with the Earth; gazing upon vast horizons that had never been witnessed by any other name-casters; reading the world like a new talking book whose pages were just waiting to be turned, whose sentences were anxious to be read and whose names were just waiting to be pronounced.

When these cognitive, linguistic beings formed naming relationships with the ancient world and its creatures, they embarked on a transformative engagement—a playground where the form of their surroundings became sound and meaning, which could then dance through the matrix of their ideas and intentions. To name something was to possess it in the body of the known and thus enable the namer to work with its identity. According to the holy Koran it was by way of this naming that the humans assumed dominance even over the angels (despite the angel's reservations on the matter).[13]

From where we sit now this prospect sounds frightening—having the fate of all living things at the mercy of linguistic associations invented by these clever and reckless creatures. But we are viewing the world from our modern times, when all named Things have suffered and endured many—perhaps hundreds of names by now; all not quite understood, all calling for another shot at "ringing that bell" and sounding their true identity.

The first namers may have recognized the importance of their sacred task and weighed this responsibility into their acts of speaking. If giving voice to something was a way of manifesting it into reality, then speaking a name was itself a manifesting action. This relationship between word and reality is represented throughout the histories of civilization. A good example is found in the beliefs of the ancient Egyptians, wherein the tongue was considered the steering pole[14] or "oar" of the soul; the path-maker that set the course of action.

In the theology of the ancient Egyptians, the complete person constituted up to nine elemental parts. These parts included the "Body, the Mind, and Soul," prefiguring a common Judeo-Christian tri-partite concept of the "whole person," but they also included the Heart, the Shadow, the Personality, the Spirit, the Power, and the Name.[15] Of all of these, it was the name alone that could be lost, transferred, or conveyed to others. It carried such importance that no creature, place, or inanimate thing could be said to have an existence until it was named.[16] A person possessing the true Name of another could wield power over them.[17]

Remembering one's true Name was also prerequisite for entry into the afterworld. "Giving mouth" to the deceased was a solemn and involved procedure that allowed the desire of the heart to enter the halls of judgment. Once the heart was weighed on the tongue of the balance of life, the Name was entered into the register and they were allowed to continue on their journey.[18] These ancient Egyptians apparently beheld sound as a sacred substance and affecter, believing that names contained forces that could direct, influence, heal, create, and destroy; and that without a tongue, a person was like a boat without a tiller.

The "Name" infers participation with others; only the Gods could self-create. Osiris "…brought my own name into my own mouth,"[19] all other mortals needed someone to give the name—someone to know the name, and someone to behold it. True names were powerful entities, dispensed with under special circumstances to convey their power, held in secret by the namer and the named.[20]

In consideration of this, the Word was not just a simple tool to be bandied about to represent something alive or sacred, the Word itself was alive; it was sacred. Speaking invited the exchange of possibilities. The voice was intention. Uttering these "words for things" was more than developing a representational vocabulary; rather it was more akin to setting things in motion. David Abram, in his book "*Spell of the Sensuous*" wraps experience and cognition around the phenomena of language and engagement, proposing that the impact of words is greater than their actual "meaning". He argues that while words do denote specific things, feelings, or actions, it is the "sensuous, gestural significance of spoken sounds—their direct bodily resonance—that make communication possible at all…the soundful influence of spoken words upon the sensing body—that supports the more abstract and conventional meanings that we assign to those words."[21] This suggests that our species' linguistic skill is not merely a tool for conveying serial information to others who share our vocabulary, rather it is a biological adaptation, like the direction-giving dance of the bees—for weaving ourselves into our environment.

This idea of the "living word" may seem mere poetic speculation, but consider the relative impact of shouting "FIRE!" in a crowded theater situation. This benchmark in the "free speech" discussion is considered illegal because the act of shouting "FIRE!" reaches behind the reason of the listeners, biologically transforming their bodies into "panic machines" to the degree of endangering their own lives. If the huge graphic of the word "FIRE!" was displayed on the theater screen, the message would be somewhat ambiguous—unless it was accompanied by some extreme sound. Imagine what effect this projected graphic would produce if accompanied by the music of Vivaldi or the sound of church bells tolling—or maybe even the smell of smoke.

Humans are sound-specialist animals; sounds play a central role in crafting our relationship to our surroundings.[22] We use sound to get things done—from hunting actions to boundary setting, from courtship to nurturing. We navigate these actions with utterances. When we—like the Ancients, call something a name, it is a testimonial to the inclusion of that thing into our realm. When they invoked names, it wove the named into an indelible bond of words and commitment to action. From a common understanding of the gravity and utility of sound, blessings and curses would set the courses and destinies of families, tribes, and nations[23]; the invocation of names could summon the Angels or bring on calamity. For the Ancients, names were more than just memory devices used to navigate places and recall experiences of persons and things; names were acoustical symbols—sonic talismans that conveyed the power of legacy and recognition of the named things:

> Among all the varied formulations of the First and Supreme Principal, none recurs more constantly throughout the later Vedic texts then the *brahman*. The oldest meaning of this word seems to be "holy knowledge," "sacred utterance," or (what to primitive man is the same thing) its concrete expression, "hymn" or "incantation." Any holy, mystic utterance is *brahman*. But from the point of view of those times, this definition implies far more than it would suggest to our minds. The spoken word had a mysterious, supernatural power; it contained within itself the essence of the thing denoted. To "know the *name*" of anything was to control the thing. The *word* means wisdom, knowledge; and knowledge, as we have seen was (magic) power. So *brahman*, the "holy word," soon came to mean the mystic power inherent in the holy word. (Franklin Edgerton, "The Bhagavad-Gita translated and interpreted by Franklin Edgerton," 1972, Harvard University Press, Cambridge, MA, p.116)

Working back from "language" to "word," into "sound," the "Name" is imbedded into the named, ready to be set in play—by being recognized or identified by a naming being.

I was speaking with an Eyak tribal member from Prince William, Alaska. She told me that like many indigenous languages, Eyak has a diminishing number of Native speakers. Until recently in their nation there was only one Grandmother who spoke Eyak as her primary language.[24] Some folks are attempting to blow on the embers of their language to keep it alive—teaching the kids; writing down the vocabulary; learning the syntax and grammar, and recording the truncated conversations. The Grandmother was not so concerned with the demise of Eyak language. Her advice was to learn how to speak a kindred language from the same linguistic family such as Diné, which has a very different vocabulary born out of a different landscape, but is nonetheless a vital and healthy Athabaskan dialect.[25] Once the dialect is learned, the perceptual framework of language can then be brought home and used to express the local experience. She said that as long as the land exists, the words will return.

The Persona

If the names are born out of the earth, the pronunciation of these names requires an interlocutor; a body to engage in the experience—inhaling the essence of the spirit on the wind, exhaling an enchanting voice to express the relationship—through sound—*per sona*. Without a tongue a person is a soul without a tiller. Without sound, a person does not exist, and our personhood depends on sounding it into our surroundings. Our task of engagement, boundary setting, coercion, and persuasion depends on how well we articulate these needs. Not through vocabulary, but through sound and inflection.

We can only speculate how the first name-makers sounded as they spoke; whether it was in the droll tones of a modern day telephone operator, or more animated, like the language of the birds. We don't have any voice recordings that antedate writing, but there are some clues as to how early speech was delivered, at least in literate times: Hamlet's instructions to his actors may be the most familiar observation on expression and delivery ("Speak the speech, I pray you, as I pronounce it to you, trippingly on the tongue...").[26] In *Rhetorica*, Aristotle also speaks in some detail about the importance of the tone, volume, meter, and rhythm of oration.[27] We can understand these texts in relation to our contemporary speech delivery, but neither of these literary works confirm how common parlance was delivered in their own times. We know that there was a difference between conversational speech and oration; otherwise there would not have been a need for the instructions.

The idea of *personae* was expressed in the theater trappings of ancient Greece. The "Personae" were the masks used to project the character and voice of the actors into the audience. Some of these theaters were quite large, seating thousands of people, most of whom were too far from the stage to see the subtlety of the actor's facial expressions, so the personae were used to visually emphasize the dominant emotional characteristics of the rôle in an exaggerated form.[28] Because the masks would otherwise cover the mouths of the actor, they were crafted to resonate and thus amplify the human voice. The mouth was always widely open or included a voice projection horn, allowing the actor to focus their voice through the mask while wearing it, animating the character with sound.[29] If these masks were visually exaggerated, perhaps the voice was exaggerated as well, accentuating even more the emotional characteristics of the rôle. If the voice inflections were exaggerated, how

Fig. 2.1 Greek Theater Mask Stoà of Attalus Museum (photo by Giovanni Dall'Orto)

deeply did this stylized manner of speaking reflect the depth of common conversational affectations?

We could gauge this in theory by looking at contemporary theater and live stage performance. The craft of theatrical acting is by nature a craft of carefully mediated exaggeration. An actor needs to work a finite space and expand or contract it through the inflections of their voice and actions to suit the scene. If it is done well, the exaggeration is hard to notice; done badly, it is just bad acting.

The craft of theatrical exaggeration is really distilled in Japanese Kabuki Theater and Rakugo (comic storytelling). Kabuki and Rakugo actors really push the envelope of expression of their character's voice for a grand theatrical effect. The acting styles are expansive and musically fun (though I don't find the embellishment too far afield from common Japanese conversational inflections). Even without understanding the language, the expressions of incredulity, deceit, passion, or embarrassment are all very clear. What is remarkable in Kabuki or Rakugo is the density and focus. A comic actor saturates their audience with an embellished delivery for the entire duration of a performance. Anyone carrying on this way out in the streets would probably be locked up.

Theater and literature are the only tangible evidence we have in determining the auditory history of common vocal expression. Unfortunately living theater is not a reliable vessel for historicity, as the rhythms and tonal inflections are mediated as much by audience responses as by artistic license. Thus the inflections mutate over time to reflect the contemporary sensibilities of the audience. On the other hand, written words more closely frame the perspective of the time they were written, which is particularly informative when the writing is traceable back to a particular author or a continuous literary tradition—such as Shakespeare, Basho, or the romantic poets of Caliphate Spain.

But prior to the convention of claimed authorship, writing was used more by storytellers to keep their story straight. And in this, the first writing forms were less linked to actual words, rather they were mnemonic devices designed to help orators track the legacy of their tales. The hieroglyphics of Egypt are representative of this, wherein pictographs emerged out of associations, which eventually became abbreviated to represent strings of sounds that could be assembled into literature or history. Contemporary Hebrew also illustrates this, in that it consists of consonants only (Latin: *con sonāre*, "with sound"). Reading silently from the page yields nonsense strings of letters; in order for it to make sense it needs to be read aloud.[30] (To some folks this manifests in the belief that the vowels are the spirit of the word, and thus consider it blasphemy to write the vowel into the word G_d.)

The Greek alphabet transformed this consideration with the inclusion of vowels (from Old French *vouel*, "giving voice"), permitting a solitary reader the luxury of dwelling on the page with the words, uninterrupted by the need to speak. Even with the addition of this luxury, throughout various times silent reading has been considered everything from impolite, to strange, to sacrilegious. Prior to the spread of common literacy, watching someone read silently must have been uncanny—or even worrisome.[31] Reading silently transported a reader into an imaginary realm which they inhabited apart from those otherwise inhabiting their surroundings, inducing a form of madness. (It is this form of madness that afflicted Don Quixote, who Cervantes juxtaposed against the "wise" orality of Sancho Panza, his illiterate and apothegmatic sidekick.)[32] Perhaps one of the symptoms of this madness was the way reading silently transformed the way people spoke with each other. The imagined "sound" of written language can be self-mediated by the reader within the silences of their own mind; too much of this inward dwelling without speaking the words produces an odd sounding person.

I know this from my own experience. When I graduated from high school, instead of immediately heading off to foreign lands or to university, as did my peers, I headed out into seclusion in a small cabin in the woods. Loaded with mountains of books from a summertime job in a bookstore, I spent a good amount of my time reading. Unmediated by human contact or sensible conversations, I silently soaked up fresh ideas and new vocabulary exclusively from books. Every few weeks I would return to civilization for conversation (and to do my laundry). Folks indicated that I was hard to understand, and they needed to correct my pronunciation quite a bit.

Perhaps there is some form of this madness inherent in the reading practices of literate cultures. Barry Sanders, in his book "*A is for Ox*" suggests that the ability to dwell in the silent consideration of written ideas forms a perspective of an inner self that does not exist in oral-dominant cultures. Self reflective critical thinking is unique to those who can read and reread a composed sentence to derive or construct meaning out of it. The "sounds" of the sentence are a co-creation of the author and the reader only, elaborated internally to support a fabricated reality unique to the reader.[33] The reader can then bring these novel perspectives into the world, untempered by community discourse.

This proposal may sound absurd to someone so naturally reading this book, until you ponder the fact that among the 3,500 spoken languages at play in the world today only about 75 are literate, and that the balance of the world's oral cultures are increasingly subject to the whimsy and wiles of the speakers of these few literate languages.[34] (Consider also the etymology of the word "absurd" derived from the Latin *ab surdus*—"unheard.")

A "post modern" outgrowth of the internal literate experience is our broadcast media (theater in a dinky box), which has further modified the way we speak. We do have auditory records of this, and these records may give us some hints as to how the range of vocal expression has mutated over the brief time since the invention of sound recording. I am thinking here about "news casting" voices and the sullen gravity of Edward R. Morrow, or the paternal authority of Walter Cronkite—voices of the evening news in the 1950s and 1960s respectively. These voices worked well in their times, but I suspect that their rhythms would be too slow and their sense of drama too "thoughtful" to cut though the dazzle-haze of contemporary media where newscasters and pundits continuously break into each other's sentences.

While the inner landscapes of reading or the framed external theaters of produced media have affected the sounds of our words and speech, I don't think that these phenomena have changed the fundamental meaning of sounds. What remains constant over the history of human communication are the paralinguistic cues of tone, volume, meter and rhythm, which are much more evocative than a concise vocabulary. It is not the words, but how you say them that deliver the juice. People will respond predictably when yelled at; it will startle or alarm someone even if the words being yelled are nonsense. Yelling is a very blunt tonal tool and it always works (yelling faster works faster). Of course we have far more delicate tonal tools at our disposal for crafting the subtle expressions of our desire—a sensitivity that extends way beyond the mere down-lilting of a word to express disappointment, or the slight ascending of pitch at the end of a question. The rich information imbedded in the tonal stresses of our sentences may even evade conscious recognition, but nonetheless they frame our impression of what is being communicated. Tonal cues give us "hunches" or feelings of whether someone is being sincere, sarcastic, cold, welcoming, bitter, or disengaged. These subtle aspects of vocal tone are understood well enough that common computerized voice analysis tools exists that can differentiate through vocal stresses if a person is lying or telling the truth[35]—a mechanical task that can be challenging in a live, personal encounter due to the potential conflicting evidence of "honest eyes," straightforward body language, and a mouthful of deceitful, but sincere sounding words.

Divining the essence of the meaning of linguistic sounds, human behaviorist Fritz Pearls used a technique with his protégés that would immediately reveal the intentions behind their expression. If someone was complaining about something, for example, he would ask them to replace words with nonsense syllables and express their ideas through sounds only. Without the obfuscation of words, it would be very easy to hear if the complainer was whining, angry, fearful, or a just expressing reasonable objection to a situation.[36] Pearls' exercise clearly illustrates that the tone of a person's voice can convey more meaning than their words do.

Any pet owner knows this when they call, scold, or encourage their animal. Arbitrary words can be substituted for the pet's name; your little kitty "Snowpaws" will respond with the same guilty resentment to the name "Bar-stool" spoken in a scolding tone; if you beckon your dog "Storm" with the name "Artichoke," he will likely respond with the same bounding enthusiasm.

Of course the reflexive aspect of this interspecies communication is that domestic animals can use linguistic sound tools as well—in controlling the behavior of their human companions. Animal behavioral researcher Nicholas Nicastro recently realized that domestic cats use an extensive lexicon of meows, flutters, and squeaks with humans that they otherwise do not use with their kitty colleagues. Feline co-species vocabulary usually involves only hissing, spitting, yowling, and purring—basic expressions of territory and acceptance. It seems that cats are quite aware of their own species' resistance to all but the most basic persuasions. On the other hand, cats use a rich mélange of musical sounds on humans, indicating a much more complex relationship of interdependence and control.[37]

It is through sound that we convey the wealth of information to others about who we are, what we want and how we feel, a characteristic that is consistent for pretty much all sound producing animals. Sound production is a means whereby creatures may rapidly modify their disposition without changing attire. In lieu of unfolding feathers, flushing complexions, modifying pigments, or even getting up into action, a single sound may unambiguously express the intent of the sound maker, immediately transforming their surroundings by announcing participation. The breadth of expression from inquiry to rage, apathy to passion can happen as fast as it takes the sound to unfurl. We make sounds that can affect incredible changes on our surroundings without our having to touch anything. We can keep people and other creatures at a distance, or lure them closer. We can induce enthusiasm and joy, dread, or fear. We direct our sounds to assure ourselves of the dimensions, texture, and density of our boundaries; we can let other people and animals know how large we are and how much territory we occupy—and how we place others in our realm of auditory influence. Through sound we create our own universe.

Bells and Boundaries

A man is fishing on the lake—well sort of fishing; actually his pole is fishing, he's taking a nap. But in the chance event that some hungry fish takes an interest in his hook, his pole will let him know, for at the tip of the pole the man has attached a bell—now lightly suspended at the perimeter of his consciousness; standing sentry between the fate of a fish and the landscape of the man's dreams.

Bells are often found at these human intersections—helping define the boundaries and perimeters of our interests. We know how bells mark the divisions of time, metering the flow of events, but they also mark boundaries of space, the extents of territories, and the reach of will. Ranging from the 375,000 lb "Trotskoi" bell in the Kremlin, Moscow [38] to the snuff-can jingles on the jingle dress of an Ojibwa dancer,[39] bells have served to set boundaries and attract attention, keeping bad spirits away, and calling in the community. Their role in society has been secured by the fact that unlike other musical instruments which need to be blown, plucked, fingered and navigated, bells are self contained—given a little motion they play themselves.

The archeological records indicate that the "crotal" or "jingle" bell was likely the first bell type. Derived from a seed shaking in a dried pod, these original bells were fabricated out of wood or clay, and eventually from metal. Hung around the necks of turkeys, chickens, goats, monkeys, cattle, and other domesticated livestock—animal bells served the dual purpose of helping an owner locate their foraging animals, while the alien, unnatural sound of the bell would ward off predators.[40] Similar bells adorning a dancing body would yield auditory feedback on how the body was moving—expanding the inhabited realm of the dancer out into the range of sound. So from the earliest records, the bell has served as both a perimeter and boundary setting tool, and an attracting or locating instrument.

The first cast-metal bells appeared in China c. 1700–2000 B.C.E. Legend of their use includes employing bells as long distance alarms or communication devices,[41] warning of encroachment, calling in the spirits, and establishing the sphere of influence of the bell-sounding people. Through the history of Christianity, bells have served in this same manner; informing the community of various events and establishing inclusion for all those within hearing distance: It has been a long standing tradition that a Christian parish was defined by the reach of the bells.[42]

The first bells of the early Christians were decidedly manual; they were portable hand bells that helped orient the faithful to each other in the deserts of Egypt.[43] Originally just calling people in to prayer, the church bell eventually evolved into an announcing tool, marking events within the day associated with the prayer times or "offices," dividing the days and the passage of time into transitions marked by sound.[44] Prior to the mechanization of time through clocks, these temporal divisions were driven by complex interdependencies of season, weather, agrarian work schedules, holy days, sleep requirements, Roman convention, the Rule of Benedict, and the metabolism of the local priest.[45]

When any population was served by a single church this worked well; all citizens were loosely synchronized to each other's circadian rhythms, tattooed by the sounds from the bell tower. But as populations grew throughout Europe, more churches, chapels, cathedrals and abbeys intersected each other's acoustical space. The Middle Ages in Europe saw such a stunning acceleration in church building that by the early fourteenth century there was a church or chapel for every 200 inhabitants.[46] With each institution ringing in their own schedules, any city soundscape must have sounded somewhat like a continuous carillon.

The introduction of the clock didn't improve things as one might expect, because the clock allowed the automatic chiming of bells without human intervention. Clock bells didn't necessarily replace the "qualitative time" rung by the church; rather it introduced "quantitative time" into the public narrative.[47] This allowed any civic institution with an interest in time to have it expressed in sound, unhinged from the temporal laws of the church. Clocks also allowed for the ringing to be staggered, offsetting any set of chimes from adjacent bells so that each set would not be masked by other bells ringing at the same time.

As cacophonous and goofy as this sounds, it is a situation that still exists in some places. On a recent trip though Mexico I stayed in a hotel on the waterfront in Mazatlan. The hotel was surprisingly inexpensive; it was only when I turned in for the night that I found out why. It was located between a small church and the harbor Customs House. Both institutions had their own clock, each with a set of bells that would ring the hours and the half hours. By some devilish agreement they both decided to disregard the convention of Greenwich Mean Time and offset their bell sets by 15 minutes to avoid overlap. Fifteen minutes is just about enough time for me to drift lightly to the perimeter of my pool of dreams... but not quite. Just as I would arrive at the gates of sleep, a peal of bells would ring out announcing either chapel time or mercantile time. I had many, many conscious thoughts throughout that night, mostly about bells, commerce, the church, and the mechanization of time. Few of these thoughts were nice...

The long association of bells with Christianity is due to Christian's extensive use of bells throughout history, but it is also likely due to how readily Christian theology resonates with the various bell metaphors: The single clapper striking the hollow metal shell—bringing it to life with a pure tone; the reach of this sound out into the surrounding void; pushing back chaos with a beautiful ringing; and calling the lost in from the wilderness. From this perspective, the sound of the bells conveyed doctrine—resonating associations with faith and spirituality.

This experience of "doctrine through sound" may seem hard to grasp from where we dwell in our modern soundscapes of rubber, concrete, and steel: surrounded by the sounds of road traffic and airplanes flying overhead; masked by the complex electronic sounds of media, and buried by our ability to produce our own deafening noises—the sound of a ringing metal bell does not seem that remarkable to us. But in the soft dirt, wood, and mud soundscapes of the fourth century holy lands[48] (when the Christian bells first sounded), the ring of a bell must have been a stunning clarion.

Empowered with the novelty of this auditory sensation, bells were easy to hear in pre-industrial soundscapes. But defining a parish as "that which is within auditory reach of the bell tower" did not end with the mere functionality of hearing it from afar. Bells invoked deep emotions, for which they were honored and thus played into a deeper sense of community belonging. The spirits of the bells were sanctified with name, purpose, and intention. They were adopted by godparents[49] and lovingly hung in amongst the family of bells that defined both the perimeter and the heart of a community, ringing out the collective sensibilities and character of the inhabitants. Given names such as *vivos voco* (I call the living), *defunctos ploro* (I mourn the dead), *pestum fugo* (I drive off the plague), and *fulgorem frango* (I break the storm clouds) indicated the various attributes of (and faith in) the power of bells.[50] These bells were more than functional beacons conveying messages from the bell towers; they were the sound of the parish identity. Welcomed in as family members, nurtured and celebrated, the bells served as voices for the community's priorities and beliefs.

As the legacy of bells thickened over time, their complex meanings thoroughly embellished the soundscapes which they framed. The usefulness of a bell to call or alert the public—and stir their emotions—was not lost on the civic minded. Sponsored by merchants, luminaries, and trade organizations, the bells were increasingly employed for secular purposes; calling town meetings, announcing births and marriages, the opening of markets, warning of hostile encroachments, calling citizens to arms, welcoming ships and wayfarers, tolling deaths, expressing joy, signaling rest, or commemorating freedom.

Given the power of the bells to express the priorities and emotions of those who invested their faith in them, the control of bell ringing has always been a matter of contention. While the church steeple was the bell's home, the clergy didn't always control the ropes. The fuzzy boundary between the secular and the sacred was smeared around community signals of alarm, harvest, celebration, and honorary peals for civic occasions. By holding the rope of the bell, the ringer could make a statement that would set the community in motion; informed by the legacy of the bell along the course of community events. The holder of the rope held the heartstrings of the community, thickening the history of bells and boundaries with copious accounts of political intrigue, deceit, revolt, and will.

The control of the bell during the French revolution was really emblematic of this. It was through this time from 1792 through 1806 in France that the Church was ripped asunder between Royalty, the Republican Revolutionaries, and the emotional and spiritual landscapes of the citizens. In this era, thousands upon thousands of bells were silenced, abducted, dismounted and hidden—or "captured" and melted down into cannons. Once the revolutionary smoke had cleared, the return of the remaining bells was itself a painful process. Parishes and hamlets collapsed into each other, abducted bells from the center of a previous community were relocated to the centers of others, and local, regional, and national laws outlining the permissible uses of bells fueled feuds and deathly hard feelings for decades afterwards. When the bells were restored to the soundscape, their legacies and original meanings had been compromised, and their use in turn shifted to reflect a more secular society.[51]

Defining the center of community and the perimeter of society with sound was not limited to Christianity and surely antedates the earliest metal bells. When journalist Henry Stanley went into Africa to find Dr. David Livingstone, he was surprised that the natives knew of his arrival well ahead of time, due to the "jungle telegraph."[53] And while there is no written record verifying the history of the "jungle telegraph" drums in Africa, it is likely that the earliest known evidence of toolmaking hominids in eastern Africa carried with them the seeds of this form of communication.

The tradition of broadcasting community news on drums is so ubiquitous in West Africa that contemporary short-wave radio stations still use log-drum "call letters" as station identification,[54] expanding the historic practice of keeping the news of tribal settlements in the air with a continuous tattoo of rhythms and defining the sphere of their common influence.

The sound of the split-log drum is low and large, producing a deep infrasonic energy that at a distance is more felt than heard, and with a distance-penetrating quality that is quite handy in communicating through densely foliated areas. When entering into the realm of the drums, early European travelers and missionaries often spoke about how the infrasonic pulse produced feelings of vertigo or anxiety—even while they were unable to distinguish any beating rhythms.[55] While the jungle telegraph is most often associated with the jungles of Africa, I was not surprised when I encountered some ancient log drums in the jungles of Yucatan—said to be used for long distance communication.[56]

This early form of long distance communication could be looked at from the standpoint of our telephone paradigm—that when some news of interest pops up, a drummer would saunter over to the telegraph drums and blow out a few lines. But unlike the telephone, everyone within earshot of the drum was in on the news. In this sense the jungle telegraph is akin to the alpine yodeling of central Europe—a form of overland communication that was common until after WWII. Due to the huge up and down vertical efforts required to travel between the mountain villages (which may have been horizontally separated by only a few miles) pre-telephone inhabitants used yodeling to convey information across the deep valleys. While yodeling didn't have a complicated lexicon, it could clearly convey emotions.

A Moldavian Elder told me that all of the villagers could understand the meaning of the yodel due to the emotional feel of it, so occasions of marriage and birth, illness and death were all understood by the emotions conveyed through the tone of the yodel. Everyone already knew who was courting, who was pregnant, and who was frail as a matter of course; the yodel would set folks into action with invitations to weddings and christenings, or a call to prayers or funerals—conveying the common emotions that the community felt around any particular situation.

In Islam, it is the call of the Muezzin from the minaret that defines the community reach. The original tale of the first Muezzin describes a disciple of Mohammed who had a dream that he encountered a man carrying a large bell. He offered to buy the bell so that he could "use it to call the people to prayer, as the Christians did." The man said "would it be better that I taught you to sing so that you could call the people to prayer anywhere?" When the disciple awoke, he told his dream to Mohammed. The master told the disciple to tell the dream to the Abyssinian, Bilal—"…because he has a better voice than you." The disciple did, and Bilal became the first Muezzin, a black man from Sub-Saharan Africa.

Every Mosque has at least one minaret from which the call to prayer resounds. In its original setting this established a Mosque's sphere of influence to the reach of the Muezzin's voice. Their call to prayer is unambiguous; it occurs predictably five times a day, it serves a singular and clear purpose, though singing it requires a gift of voice. One of the beauties of this is that the call of the Muezzin cannot be hijacked to serve any other political or civil purpose. If someone other than the Muezzin was up in the minaret, it would be pretty clear who they were and what they were up to—unlike the chiming of bells, whose ropes can be pulled by anyone from below.

Fig. 2.2 Bell Yard in Hamburg, Germany[52] (Percival Price collection courtesy of Library and Archives Canada)

The call of the Muezzin is also more centripetal than bells—it is an attracting sound rather than a boundary-setting sound. This characteristic probably has much to do with the desert geography of early Islam, were there remains a high degree of nomadicism to this day. The minaret that we associate with the Mosque predates Islam and served as lighthouses in the horizon; signal towers and beacons welcoming travelers into the hospitality of the caravanserais.[57]

The "reaching out and awakening" of the Christian bells, and the "calling in and welcome" of the Moslem Muezzin still serve as dominant metaphors of these two western religions. The Jews on the other hand have not cultivated an evangelical out-reaching practice nor an "attracting" architecture: Towers and clarion calls are counterproductive to a persecuted people. Though somewhere between the centrifugal idea of the "bells as messenger," and the centripetal idea of the Muezzin's "call to prayer" is the Jewish *Shofar*, a ram's horn blown for ceremonial purposes. The Shofar is not particularly musical and is not played for pleasure; rather it is a signaling device used variously in the contexts of waking the people up from spiritual slumber, overcoming the forces of evil, and getting God's attention.[58] The Shofar also wields supernatural or sacred power. In what is probably the most spectacular uses of sound to get something done, the Shofar was employed as an instrument of war by Joshua to bring down the walls of Jericho. Laying siege on the city, Joshua instructed seven priests to compass the city for 7 days blowing seven "trumpets of rams' horns."[59] The Shofar is not a pleasant sounding instrument anyway, but to the inhabitants of Jericho that week, it probably sounded really lousy.

All of the above mentioned acoustical boundary setting behaviors are in social and societal contexts which reflect and sustain a myriad of ways that smaller social groupings—tribes, families, and especially individuals, set their own acoustical territories. This is a human characteristic that has remained with us throughout time. But if there is a new textural thrust to modernity, it is the ever-increasing dominance of the individual in the social milieu. We are less subject to the societal constraints of time and territory than we were even 25 years ago. The advantage to this is that we have more personal "freedom" than in previous eras; the downside is that the reliable boundaries of social convention are now left up to each individual to monitor on their own. Dinner time is no longer at 6-o'clock; the neighborhood no longer rises with the sun or goes to church each Sunday morning. We can do these things if and when we choose, but these personal choices require us to administer the actions ourselves, without the encouragement of everyone else doing the same thing at the same time.

To help with administrating our individual tasks, we have mechanized many of them. Alarm clocks, "feeding schedules," and "play dates" are all artifacts of this. With the perimeter of our community no longer defined by city walls or the sound sphere of the bell tower, individuals are left to mediate their own acoustic territories, which we have mechanized as well—and not necessarily in a refined manner. The artifacts of this are becoming ever more contentious in modern societies; the car horn (and its idiot bastard son, the car alarm), loud exhaust manifolds on motorcycles, louder and more complex police sirens, bull horns, and behemoth car stereos— all pushing out an individual's acoustic territory into the territory of other people's silence.[60] This is particularly evident in America, where the cult of the individual can disproportionately supercede basic civility.

The invasion of public soundspace by personal noise is not necessarily driven by technology; rather it is driven by a need to define personal space in our modern society. This was made particularly evident to me on a recent trip to Egypt. I was told that modern Moslem cities are very noisy, and that the sounds of technology blare throughout the day and night. I assumed that western technology (having been just dropped into these ancient cities by Europeans sometime during the last half century), was handled recklessly by people who had not developed our modern sensibilities around it. Reckless handling can be the case around the community noises of loudspeakers and industry, but surprisingly it was not the case around Egyptian personal noise.

Car horns, which I did hear continuously—even throughout the night, are used in a completely different manner than we use them in America (where the car horn is most often a personal extension of anger or anxiety). The roads in Egypt are unlike the tracks of destination found in the west; they are rather like paved "tendencies" strewn with potholes and the detritus of opportunistic use (extracting palm fibers or drying fish on the road beds, for example). As roads they are used for all manner of traffic, simultaneously displaying a 6,000 year history of transportation—from walking and goat herding to pack camels and donkey carts—all intermingled with motorized vehicles of all stripes. The cars and trucks are not expressions of personality here; rather they are more like motorized beasts of burden.

In this chaotic setting, driving in "traffic lanes" is useless, so the drivers weave toward their destination using the car horn as a courtesy signal, notifying slower traffic of a rear approach with a delicate "tap, tap" on the horn. This tapping is so habitual that even in cases where there is little risk of collision, the "tap, tap" is still expressed, sort of like a "tipping of the hat" to other drivers. Hundreds, or even thousands of these horn taps calling across the city soundscape does thicken the noise field considerably, but it is not the angry sound of car horns heard in the States, rather it is the sound of courtesy—altogether a different thing.

These western cultural perspectives on how humans sound out our sphere of influence only hint at the rich legacies found in non-western and indigenous sound-play at the perimeter of their respective societies. The rattles and bones of African Griot, the drum of Native American Medicine Man, the chants and whistles of the Sammi Shaman, and the bells and horns of Tibetan mystical healer—all use sound as boundaries and gateways between the human world and the spirit world in ways that are equally compelling and just as complex.[61] This all points to a more fundamental characteristic of our species; that creating acoustical territories may be among the first expressions of human will—second only to the will to emerge from the womb.

Taking Control

An infant's first draft of air expresses itself in sound—a cry with the power to buckle the knees of the strongest man and to loosen a flood of tears of all those within its acoustical realm. African Shaman Malidoma Somé tells us that when a child was born in his village, all of the village children would wait outside of the birthing hut. When the child issues his first cry of breath, all of the children of the village yell back in welcoming affirmation.[62] In Somé's culture, the infant's first cry is reinforced by community response. But even in less intimate societies such as ours, a child learns at an early age about the power of personal sound—how sound can beckon or repel, and how sound can get everyone moving.

From the first responses we stimulate in others with our sounds we learn quickly that we can produce an impact on our surroundings without our having to touch it.

Our voice precedes us as we grow, helping us keep other people and species at a distance, or lure them closer. Through the sounds we make we assure ourselves of the dimensions, texture and density of our physical boundaries, and we can let other people and creatures outside of our visual realm know where and how large we are.

Infants are capable of expressing huge amounts of emotional information almost immediately after drafting their first breath of air, and once an infant gets a grip on their providers by way of sound, they continuously test it. They test it under varying circumstances to express hunger, solitude, physical discomfort, desire, and pleasure. Soon a common "pre-linguistic" vocabulary is established between infants and their providers with the underlying premise of "I make sound, you guys move." The whimpers, sighs, grunts, and coos of a newborn child have the ability to summon deep emotional responses in most other sentient beings. Their sounds are so compelling that they can strongly affect all creatures within earshot, regardless of species. Powerful infant sounds are common to most breathing creatures; whimpering puppies, mewing lion cubs, whinnying foals, cheeping chicks, even a salamander pup's squeaking are sounds that can passionately motivate adult animals across species lines.

Infant humans realize this almost immediately; what they can't touch with their hands, they move with sound. Their first vocabulary consists of a panoply of sounds that affect others: fawning, cooing, gurgling, screaming, crying, whining, pouting, yelling, singing, even silences are used well before an infant understands symbolic or representational vocabulary. With these sound tools, the child manages to convey enough information to tailor their care and modify their surroundings for their comfort. Even as they learn words to convey ideas, they rely on the tonal vocabulary of sounds to convey those words because the sounds are stronger emotional motivators than the words alone. We continue to craft and refine our sonic vocabulary as we become more articulate, and while we may understand a conversation from the perspective of sharing ideas through words, the sound qualities we make in conversation are likely to convey the more important information—information about how we feel.

Whether we know it or not, we refine the inflections and textures of our expressions to more accurately achieve our desired results, tailored by the responses we get from others. By the time we reach adulthood, our vocabulary of subtle sound inflection is so rich that we can easily identify each other in the smallest snippet of sound. This partially accounts for our ability to recognize an unannounced phone caller on their first "hello"—even if we haven't heard from them in years.

Subtle voice inflection so persuasively conveys personality that it might as well just be a name tag. Though unlike a name tag, or even a face, human sounds hook us into behavior. Sound conveys emotion so effectively that hearing a familiar voice immediately establishes an emotional relationship—not just a spatial or temporal one. Whether we are cognizant of it or not, most of our memory cues of comfort or suspicion, mistrust or safety—come bundled with the familiar sound of someone's presence.

Using tools that allow us to dominate our own acoustic surroundings, we also protect ourselves from each other's invasive sounds. We do this by insulating ourselves against sound intrusion—building our surroundings to exclude outside sound—or by isolating ourselves from other people's sound-spaces by withdrawal. We can distance or isolate ourselves by "masking," or creating personal sound spaces that are louder than our surrounding soundscape. The instruments we use are the personal automobile, the personal headphone, the personal work station, the personal entertainment center, the personalized internet browser. With these tools, and the accepted social conventions that allow us to withdraw from visceral and auditory contact with each other, we are left in complete control of our personal acoustic environment. Mediated by our own will, it becomes hard to gauge the scope of any social or community interaction. Increasingly our only "community" feedback is through pre-produced sound sources—the radio or television. In lieu of dwelling within earshot of the minaret or belfry, we increasingly dwell within a personal acoustical perimeter of our own design. By taking personal control of our soundscape, we may be losing contact with our human society.

This is particularly poignant right now, because as I write this the U.S. 164th Marines are sitting in the middle of Baghdad picking the bones of conquest out of their teeth. The millions of people who assembled across the globe attempting to stop this action were dismissed as a "focus group." Away from the plazas and parks that contained the rallies, the sounds of political unity were not heard—stonewalled by a fragmented, media-driven and screen-focused society. Waiting to hear their own voices on the radio and television, but largely silenced by exclusion; only having the polished voices of celebrity newscasters to convey some minor details of their urgent message.

So when millions of unified voices are silenced by media exclusion, the "political rally" is perhaps more useful as a unifying force for the participants, and not so much a vehicle to express ideas to a largely dumb media machine. The advent of a highjacked media actually clarifies the importance of gathering together, breathing common air, and sounding out—and listening to the scope and scale of our community noise.

Community "sounding out" was successfully used in the recent Serbian struggle for democracy. The Serbian protesters and their supporters identified themselves to each other by using whistles, which they blew to drown out the distorted state radio and television coverage. When the state television news began at 7:30 p.m., a cacophony of whistles erupted, accompanied by sympathizers beating on garbage pails and sauce pans.[63] The noise was so pervasive across Belgrade and Kosovo that it was undeniable, giving further confidence to the citizens in their fight for democracy.

Community sound is a great political unifier, to the degree that the form the sound takes may not be too important, just as long as it is a common sound—a sound that can be joined. In 1992 I attended a San Francisco political rally on the occasion of International Women's Day. I usually avoid these things because I don't really enjoy huge crowds. This event was different for me because there were some issues coming up to a congressional vote for which I felt strong enough to show my public support. Besides, as a single male I couldn't help realize that there would be other opportunities present at a congregation of 100,000 or more people—only 10 % of which would probably be other men. So I packed my political and mercenary self onto the ferry and shipped out across the bay to City Hall.

The rally was taking place on a Sunday so there was very little traffic or building noise downtown. From the ferry docks there was a march along Market Street for some distance, and the sounds of many small pockets of sloganeering in the huge whooshing sea of mostly feminine speaking voices was itself a unique soundscape, but what I found truly unique for my ears was the circus atmosphere of the rally itself, with various orators and political speakers all trying to whip the crowd into a vociferous frenzy.

To be fair, there were actually some good speakers present—Pat MacDonald, then the president of the National Organization of Women presented a cogent and informative speech, as did Norma McCorvey, of Roe v. Wade fame,[64] but by-in-large the speakers who said less got a bigger rise out of the crowd. By spinning digestible, rhythmic phrases out, they could capture the mob's desire to get pumped up, to speak—or rather yell—in one voice. California Congresswoman Nancy Pelosi was just in her first term at the time,[65] and while I could see that she was not afraid to invite the crowd to join her in some serious sloganeering, her rhythm, conviction or delivery needed some work. The real *Grande Dame* of the rally that day was California Senator Barbara Boxer, who, as one of the few women in the Senate at the time, clearly demonstrated her senatorial skills by reducing the whole crowd into a huge pulsing slogan. I couldn't help think of the "four legs good, two legs bad" chant from George Orwell's rebellious Animal Farm.[66]

At the end of the rally, the crowd broke up and the participants melted away, exhausted, catharted, and spent with what I'm sure was a feeling of belonging that lasted long after their ears stopped ringing. I realized then that these rallies are not really a good forum to feed a quest for information, rather they are a chance for a crowd to lift up their voices as one, to be heard, and to let the media, their city, and their country know what everybody already knows. Without the mob behavior, the yelling of rhythmic slogans, and the beating of drums, a rally would be ineffectual. After attending, I could only imagine an equally huge rally held in complete silence. I'm sure it would be quite frightening.[67]

Of course due to the impact of these large gatherings, and the affect on unification of the spirit, the very control of these sound spaces becomes a point of contention. In the Laws of Plato's Republic, he scaled the size of an ideal "state" to 5,040 citizens[68]—a quantity of people that could be realistically addressed by a human voice. In Plato's time the possibility of this urban scale did exist—and was close to the size of many European and American cities through the nineteenth century, wherein all inhabitants were within reach of the common soundmarks, and all soundmarks were of a comprehensible human scale.

This all changed with the introduction of the loudspeaker. Perhaps one of the most profound effects of modern technology on communication is that we can now amplify and alter the sound we make electronically. We can affect the dimension and reach of words and ideas, controlling the visceral cues of scale and importance by way of electronic manipulation. The late philosopher Ivan Illich wrote that no real extreme political dynamism had occurred for centuries on the small Dalmatian island where he grew up until someone arrived with a loudspeaker.[69] This device permitted an individual to have the power to usurp the community sound space.

Adolph Hitler himself remarked in his *Manual of German Radio* that without the loudspeaker, the Nazis would have never conquered Germany.[70]

It was not only the loudspeaker that helped Hitler unify the German people. He was a talented orator, and the power of good oration is as dependant on persuasive delivery as much as it is on digestible ideas ("four legs good, two legs bad" for example).[71] Hitler was a fan of high drama, grandiose pomp and fabulous spectacle. This played well into his presentations—ably assisted by Albert Speer, the architect for the Third Reich, who employed many psychological "tricks" on the subconscious to propel the Reich into mythical proportions. At the Nuremberg Rallies, his use of heavy swastika-emblazoned felt banners did more than impress the eye when they were unfurled at the arrival of *der Fürer*.[72] The stadium's reverberant field with the loud anxious din of the crowd would suddenly become calmed as they dropped huge sound absorbing panels around the bright open stadium, embracing the crowd in a hush of felt and flannel. Feeling calm, safe and secure, the audience could open their hearts to the message of the Third Reich.

This psycho-acoustic trick notwithstanding, it was electronic technology during the Second World War that really transformed the way people perceived sound. The loudspeaker became a window to sounds from a world that existed somewhere else, in some other time. Radio studios, imaginary film sets, pre-recorded sound, and media sound production all forced a cognitive reality shift. For the first time in civilization sound originating from outside of the reach of the listener impinged on their bodies and their imagination. Through radio, a voice of authority would arrive into an intimate soundspace, bringing Theodore Roosevelt, Herbert Hoover, Bessie Smith or Adolph Hitler within reach of the average person's living room. In lieu of gathering around the fire or in the public square to listen to the elders, by the 1930s people would gather around a radio,[73] sharing their common space with uncommon people.

Sound technologies allowed for the crafting of soundspace in unusual ways, producing improbable juxtapositions of sounds designed to portray emotional states rather than actual settings. The sound of twittering birds mixed in with a lovely singing voice, or the beat of marshal music behind the assured voice of a war propagandist was very effective in capturing the participatory imaginations of listeners. Soundscape coinologist, composer, and writer R. Murray Schafer frames this phenomenon quite succinctly with his term "schizophonia"—referring to packaging and storing of sounds, and the splitting of sounds from their original context to craft composed, imaginary settings.[74]

This new reality was inaugurated through the radio. Soundspaces were constructed that replaced the village story teller and the town elders. A few folks could sit around a microphone and create a totally fictitious realm that a whole population could dwell in. The radio became the new church bells, if you will—which defined a metropolis as "all of those who lived within reach of the radio tower broadcast." Many of the same metaphors and characteristics apply; the partitioning of days into time segments, the embrace of the safe and the exclusion of the threatening—the voice of a secular God from above. There is a certain hyper-believability of the broadcast voice in a produced soundscape—one which I don't feel we will ever quite get over, if for no other reason than we can never directly question its authority, we can only turn it off. (But it goes on talking…)

Produced sound so effectively colonizes the imagination because we are not hardwired to question the authority of our own auditory perceptions. We can question the content, but questioning the acoustical cues of sincerity and intimacy requires a conscious effort. The shortcoming of radio (if you can call it a shortcoming) is that the message engages the imagination of the listeners; the listeners construct imagined bodies and landscapes around the sounds of the voice. This "drawback" is neatly addressed by the moving images of television and cinema. The word "imagination" implies an internal creation of images that inhabit a mental landscape, and sound germinates these mental images very effectively. Providing images along with the sound serves as an unyielding mold for the imagination,[75] a process that broadcast patriarch David Sarnoff introduced with the television at the 1939 World's Fair, stating: "Now we add sight to sound."[76] In our current, visually dominant society, Sarnoff's statement now seems quaint.[77]

The imaginary world behind the box and through the loudspeaker becomes our new "commons"—both a source and repository of common experience for large segments of society. Lured by the moving images, seduced and comforted by the intimacy of the voices, we feel that we somehow belong to that world. Serving simultaneously as the hearth at the center of our families, and signal pyres defining the perimeter of our society; the bell tower transformed again by technology.

Controlling this imaginary landscape has proven to be a huge boon to those handling the ropes. Like the ropes of the belfry, the handlers do not need to be known for their bells to ring, but unlike the bells, it is improbable that their control could be easily usurped by any members of the listening community.

Sound and Warfare

If the intent of warfare was dispassionately reduced to a socioeconomic function, it might be described as an engagement between two competing interests wherein an aggressor throws all of their precious resources at a defender's precious resources until one of the interests collapses. The desired objective is the economic attrition and/or defeat of spirit of the loser within the recovery envelope of the victor—to whom go the spoils. Many words have gone before this act to justify a human killing behavior that is universally decried by all societies (thou shalt not kill...) but nonetheless has been part of our humanity, even our species—as far back as time's window allows us to see.

All manner of wiles and tools have been brought into the "Theater of War," and while we mostly imagine war in terms of weapons that destroy life and property, the acts of war are only partially draped in maimed bodies, burned flesh, and pools of blood. Blowing up bridges and poisoning wells notwithstanding, a huge part of war involves the beating of chests and the rattling of sabers—making huge noises to intimidate the adversary into surrendering early on, postponing the inevitable bloodletting and destruction of property that actually serves nobody.

Preparations for war require unification of the warriors: Proud martial music, war dances, and drill sergeant's measured barks all set hearts beating to the tempo of courage. Battle cries and braggadocio inspire collective confidence in the

Fig. 2.3 Figure carrying a
bull roarer, the "Baboon-
man" rock painting from the
Brandberg, South West Africa
(sketch by J.R. Harding for J.
African Music V5n3)

mythical strengths of individuals, and the invincibility of the collective. Once the
tears of the Mothers have been shed and the troops have been assembled on the
battlefield, the real sound wars begin.

At home the "March of Time"[78] newsreels roll on, extolling the virtues of the
cause, and revealing some of the more magical weapons held by "our side;" a talis-
man from a particular deity, a robotic soldier, a truth-finding arrow, a smart guided
missile, a cloak of invisibility. Meanwhile before the battlefield the engines of war
thrum and roar. We know this sound today as the deep beating blades of supply
helicopters, and the thunderous scream of fighter jets ripping across the sky. These
are the sounds of fear. And while the latest incarnation of these sounds impress a
sense of immense power on the listening body, these types of sounds have been with
us for thousands of years.

Petroglyphs in Africa and Australia dating back 60,000 years illustrate a device
now called a "bull roarer." This device made its appearance in Europe 25,000–
15,000 years ago and was present in the "new world" prior to the European

invasion. The bull roarer is quite simple—not much more than a cartouche or blade of stone, bone or wood on a string. Holding one end of the string, the roarer is rapidly spun around in a circle. As the string twists from the spinning, the roarer begins to counter-spin on the axis of the string creating a fluttering sound. With some energy behind it, this flutter becomes a growl or shriek. This sound is generated from an area the diameter of the string length, creating a complex spatial signature, so the whole effect suggests something quite large and supernatural—an ideal sound to frighten adversaries and predators. The sound of one of these instruments can be eerie, but a whole infantry spinning them as they advance over the horizon was probably terrifying[79]—a deep but ethereal flutter and shriek that seems to issue up from the beyond. If a defending army was not familiar with them, the sound would be particularly frightening; but even if they were familiar, the sound—like the howl of an attacking beast, would nonetheless trigger fear; the welling up of an aggressive roar into the deathly pre-battle silence would naturally amplify the terror of an oncoming threat.

The fear produced by an immense sound approaching from the horizon was a tactic that brought the Pipe and Drum corps into European battlefields. Through the eighteenth century, The Scottish Highlanders produced the sound of immensity with their "Instruments of War"—the throbbing drone of the pipes and the incisive unison lines of the chanters, fiercely driven by the ripping pulse of a battalion of field snare drums. Surely a frightening din to the foe and a bolstering of the blood-pulse for the advancing army. The encouraging music of the pipers was continued through the battle, also providing solace for the fallen. The use of pipes was so effective in war that the British government punished captured Scottish pipers as arms-bearing soldiers.

Prevailing on the Scots in 1750, the British stiffly outlawed the use of pipes.[80] Twenty years later, realizing the valuable penetrating quality of the pipes in battle, the Brits employed pipers to convey battle instructions—like the bugle callers in American infantry practice. Though in true British form, these poor souls were often placed high above the scene so as to project their instructions further into the fray. No longer just a tactical tool, they were of strategic interest to an adversarial army, and thus the pipers were often shot first.

There is an often mythologized account of Scottish bravery that describes a front line of pipers leading the infantry into battle. The myth is that the riflemen advanced up to the battle line behind the pipers and were instructed not to fire until the last piper fell. Aside from being a tragic waste of musical talent, strategically this doesn't make a lot of sense. I suspect that this story was assembled from an instruction to an advancing Scottish army not to attack until the last of the British pipers had been silenced, rendering the Brits without a signal corps.

Subjecting defenders to new, previously unheard sounds works well as a terror tactic; few sensations will put a person's body on alert more thoroughly than a new sound. In a pre-metallurgical era, bells, or the din of metal clanging would give a "wood and stone" army a strong case of the heebie-jeebies.[81] Whistling arrows, Buzz Bombs, screaming missiles, throbbing bombers—were all noises that were initially novel to their victims in their respective times. Of course once the physical impacts of these weapons became known, their sounds become even more terrifying.

During the Blitz of London, the noise of the Buzz Bombs was as demoralizing as Hitler had hoped[82]—in the same manner that the pounding blades of modern attack helicopters or screaming waves of attacking fighter jets erode the spirit of any contemporary defending population. The sounds of incoming ordinance—particularly supersonic projectiles and missiles—are never heard at the target, because the sound arrives after the hit. But the surrounding grisly noise of hopeless destruction saturates the nerves and sinews of the target victims, unleashing the pummeled emotions of survival; anger, fury, rage, shock, depression, grief, hopelessness, terror, and despair—a very unhealthy diet for the human spirit.

Sonic novelty always hits high marks on the fear scale, but there will always be a place for some good old-fashioned growling, gnashing, and screaming. Such is the reputation of the Viking *Berserkers* who struck tyrannical fear in their victims with berserk behavior—to the accompaniment of their animal sounds: "Sometimes I seem to hear a bull bellowing or a dog howling, and sometimes it's like people screaming" declares a distant earwitness account from antique literature.[83] Explosions are unambiguous, and repetitive explosions are even more unambiguous. A relentless assault of BOOM! CRASH! THUD! BANG! destroys confidence. The louder these sounds are the more hopeless one feels in their field.

Sound is the most direct and unambiguous visceral measure of scale. We are accustomed to having our eyes play tricks on us, but our ears and bodies are hard to fool. The earth shaking rumble of huge battle machines, the dull, thick impulse of a distant bomb blast, or the sharp nasty crack of a high velocity projectile—leave no questions in the body about scale, scope, and danger of the noise source. If demoralizing the foe is an objective of war, the acoustical artifacts of conflict serve the mission well. So in these modern times of automated warfare where "surgical precision" is a proud catchword, it seems that the incentive behind maintaining a nuclear arsenal, or developing MOAB ("Massive Ordinance Air Blast") bombs is almost exclusively driven by the BIG BOOM, rather than any strategic value of just making the biggest possible mess with one bomb-drop.[84]

Of course the "Art of War" involves more than the loud pounding of chests or even the pounding of adversaries into oblivion. It also involves support through communication and surveillance. In this context, sound is the pavement of warfare. Prior to radio and satellite communications, interpreting sound cues was often the most effective method of determining the progress and nature of armed conflict. Field Marshals and Commanders kept their ears to the battlefield to determine the movement of troops, the presence of action, and the size and scope of conflict.

In pre-industrial times, gunshot and cannon fire could be easily heard over the background sounds of nature. The clatter of hooves and rattle of armor were distinct sounds that helped identify the centers of military activity, but even the silent movement of troops and the pitching of camouflaged tents and screens would transform the acoustical characteristics of a forest understory or an open grassy meadow with an aberrant acoustical texture that did not square with the natural landscape. A perceptive field commander could sit above a battlefield and derive an auditory sense of the battlefield; sensing hostile troop movement, the placement of his own troops, the lay of the forests, fields and streams, the prevailing winds and oncoming weather—all from how these elements tempered the soundscape. The outcomes of decisive battles have been determined by how effectively the sound cues were heard and interpreted—or missed by the field commanders.[85]

Warfare surveillance does not end at the battle lines. Getting behind the lines to know the enemy's perspective is one of the most powerful tools of war. When people of differing racial origins are fighting, infiltration is a challenge for the intruder; but when people of the same racial origin are in conflict, discriminating friend from foe is a distinct challenge for the defender. Historically some clever sound cues were brought into play to help with this discrimination—and the advantage of acousti-linguistic differences often played a role.

Using sound to distinguish outsiders form those within the fold is represented in the word "barbarian," which comes from *barboi*, the Greek word for "stammering"—or "babbling," the strange sound of those speaking a foreign language. Even when one becomes fluent in a second tongue, often the artifacts of the mother tongue remain in the form of accents—phonetic elements that do not comfortably reside in both languages. It was through this that the World War II Dutch resistance shook German spies out of their fold by maneuvering them to pronounce the name of the Dutch town of "Scheveningen." The Dutch pronunciation of the guttural "schev" was outside of the phonetic vocabulary of the Germans,[86] whose stammering would betray their mother tongue. In biblical times this same strategy was used by the Gileadites in their war with the E'phraimites. Anyone wanting to pass over the River Jordan was asked to pronounce *shibboleth* the Hebrew word for "stream." The E'phraimites pronounced the *sh* as *s*, thus revealing themselves.[87]

In biblical times field battles were like all field battles—with opposing armies throwing bodies and blood at each other, propelled by hard objects, machines, and fear. But these ancient times were also much more a time of siege. The outlands and frontiers were ruled by brigands, thieves, and opportunists, and the cities were walled citadels that provided protection for their citizens. Scarcity and greed drove conflict. With the exception of the Crusades, wars of conquest hardly even pretended to be about ideas, rather they were about one group of folks wanting what was inside the walls of someone else's city.

Invaders would lay siege to these walled cities—sometimes for years—to appropriate the goods, kill all the men, make slaves of the women and children, and control the surrounding domains. Being surrounded, defenders would have to rely on their food stores to survive. Besiegers would set up "un-civil" engineering camps surrounding the city, building large towers and catapults from which to hurl burning objects, disease-infected carcasses, and other vile projectiles into the citadel.[88] In these settings there were three ways for the invader to get behind the walls: Over the top, through the sides (or gates), or under the ground. Scaling or penetrating the walls employed obvious offensive and defensive activities (notwithstanding the Trojan Horse stunt). Going under the walls on the other hand, was surreptitious and called on some imaginative protection measures by the defenders.

Tunnels being out of sight, defenders had to rely on sound and vibration to reveal the presence of digging. Herodotus mentions a bronze worker in the besieged city of Barca (now El-Merjeh in Libya) who used the resonance of a shield laid on the ground to help locate the tunnels of the Persian intruders.[89] In a similar but more sophisticated manner, the Alexandrine architect Trypho determined the locations of invasive tunnels under the citadel walls of Apollonia (now in Albania) using hanging resonant vessels. Once the excavations were discovered, the defenders didn't take kindly to the invaders. Opening the tunnels up from above, they poured kettles of boiling water, molten pitch, human excrement and red hot sand into them.[90]

This illustrates a bit of the gooey part of warfare and the unpleasant reality that once the conflict is over, the surviving combatants and their families are left with an untidy catastrophe of rotting flesh, festering, painful wounds, and broken souls. If somehow the strategic objectives could be met without all the mess; if I could say "Bang Bang!! You're dead!" and you would fall; only to get up when Mom calls us both in for dinner, war would not be so grisly. It is perhaps from the innate humanity in all of us that the idea of "non-lethal weapons" arises. Imbedded in this idea is a desire to "have our way" in conflict, but not have to really hurt others.

Non-lethal weaponry is not a new idea. Any conflict from shouting matches to fist-to-cuffs is a practice in non-lethality. Used most often in subduing "civil unrest," the term "non-lethal" really refers to blanket methods of temporarily incapacitating an adversary without causing permanent harm. The modern arsenal of non-lethal weapons has been cultivated by folks with a twisted imagination and reads like the contents of a "fun-bag" for a mean clown: Super sticky foams, rubber bullets, water cannons, calmative chemical sprays, laxative gas, and other niceties. Because sound is a "soft energy" it is a natural for non-lethal use.

Acoustic weapons designed to incapacitate by "acoustic trauma" include low frequency or "infrasonic," and audible band devices. "Infrasonic energy" is defined as all acoustic energy below the human ability of pitch discrimination. Quantified, it is all periodic or impulse acoustic energy below 20 cycles per second—frequencies that don't impinge on the ear as much as they influence the body. Low frequency sound saturates the surroundings and is characteristic of the movement of large things. Thunder, hurricanes, tidal waves, and earthquakes all generate deep infrasonic energy, so like the rumble of an earthquake or the deep howl of a hurricane, unpredictable noise in this band naturally triggers survival anxiety in the target subjects. This low frequency sound can also shake body and bone, resonating with organs and cavities to cause vertigo or nausea.

Some of the larger bells, and the "jungle telegraph" mentioned earlier in this chapter have caused vertigo and nausea in certain listeners. Ivan Illich mentions that the bell named St. Peter in the tower of the Cologne Cathedral (at 24 tons and 3 meters wide, the largest bell on the Rhine) caused vertigo in his colleague; "…the sound of the bell was too low for the ear but not for the guts."[91] In *Village Bells*, Alain Corbin also quotes a listener; "So violently did the reverberation of all these bells agitate the air that …those listening suffered a *sort of vertigo* and minds were distracted from any other preoccupation."[92] What better way to temporarily disable a group of hostiles than to have them dizzy, barfing, and even shitting due to powerful infrasonic noise?[93]

Unfortunately (or fortunately) it takes a huge amount of energy to generate the levels of infrasonic noise required to induce these effects, limiting their practicality. The sound generators need to be large, and due to the difficulty in focusing huge low frequency noise, the perpetrators would need to have their Dramamine and bottles of Kaopectate at the ready prior to unleashing the wrath of infrasonics on any crowd.

Sounds, or rather "noise" in the range of human hearing shows more promise from a "controllable" standpoint. Loud audible noise can distract, annoy, disable, maim, or kill depending on volume, envelope, pitch, and propagation. While an entire area can be saturated with audible noise, non-target folks can stand away, wear ear protection, or leave the immediate area. These characteristics all play well into non-lethal applications. Crowd dispersal is the most obvious use, and in these dangerous times where the peaceable assembly of citizens could be construed as terrorism, I'm sure that we will increasingly hear loud audible-band noises used in this application.

Most of us are already familiar with, and have perhaps been subjected to acousti-cal personnel control by way of the ever more annoying and loud panoply of police sirens. Unlike the fire bells and hand crank sirens of yore which signaled right-of-way for police and firemen, newer emergency signal devices are designed to get our attention and to stun us into behaving. Typical volumes of 120 dB SPL a 10 yards[94] bark with acoustic authority. Our natural reaction is to immediately get out of the way—heart palpitating and in a cold sweat.

Extremely loud noise strips the mind of reason. The whole body reacts in a flight response. If flight is not possible, panic and terror take over. Even with ear protec-tion, extremely loud sounds impinge on the body,[95] pushing the sensory and survival response envelope. I have willingly entered into environments (to work) where the ambient broadband noise was above the threshold of pain.[96] In both cases I had suit-able ear protection, but nonetheless my work was significantly impaired by the effect of the noise on my body. I can only imagine being involuntarily subjected to painful, or ear damaging obnoxious noise—particularly surrounded by a crowd of panic stricken people who all have limited avenues of escape. One of the drawbacks of crowd control through acoustic trauma is that the responses can be unpredictable. Frightened victims may run for cover, but if the victims are already furious, they may not be so easily coerced.

A more predictable application of audio band noise—just shy of acoustic trauma, is acoustic harassment. Audible range "Acoustic Harassment Devices" are used on fishing nets to scare off marine mammal net-predators[97] (or ring their dinner bells, according to some fishermen), and you can flush out skunks and other vermin nest-ing under your home by playing a loud radio in their living space. A similar strategy was used a few years ago at the Vatican Embassy in Panama, when the U.S. Government wanted their errant son, Manuel Noriega, to come home. They set up a large Public Address system outside of the Embassy and played music of "The Doors," "Iron Butterfly" and other American pop groups at high volumes, 24 hours a day for days on end. This, along with other persuasive tactics eventually flushed the despot out of the embassy.[98]

In the Noriega case the Army just put up a few stacks of speakers aimed gener-ally at the Embassy. That was in 1989. Since then, U.S. Military "psych-ops" have refined their techniques in the audible band. In addition to more powerful sound systems, the material selections have been honed; the contorted sounds of screaming babies, rabbits being killed, and other unpleasant biological sounds come from one pallet.[99] The arsenal also includes "heavy metal" rock from AC/DC, Joan Jett, and Def Leppard—all played at excruciatingly high volumes. This characteristic American music produces another effect on contemporary combatants; the US Marines engaged in the conquest of Iraq are familiar with the music and even enjoy the themes while in battle (soldiers will pepper their battle accounts with criticisms of the music selections used by psych-ops[100]); meanwhile, people not accustomed to "screaming spandex voices and razor-slashed guitars" have the assault of unfamiliarity to overcome.

In more limited applications such as crowd control and "Active Denial" applications, more precision is called for. Current technologies developed under the rubric of High Intensity Directed Acoustic (HIDA) devices include the Long Range Acoustic Device (LRAD) which can focus high intensity acoustical energy into narrow beams that remain coherent at 1 km.[101] At shorter ranges of 50 meters or less, these devices can exceed the threshold of audible pain and can cause permanent hearing damage. The US Army has deployed these in Iraq and Afghanistan, and two were deployed by the New York Police Department to keep protesters at bay around the 2004 Republican National Convention.[102]

All of the acoustical weapons mentioned above are used in terrestrial, airborne sound settings, but acoustical conflict is not limited to soft human targets. Much of our military engagements occur in and around the seas, and dominance of littoral waters around "theaters of war" is a key to any international military venture. Modern global conquest requires moving aircraft carriers and their support fleets around the globe. These fleets are well protected and hard to approach from above the ocean surface, but they can be vulnerable from below. If a submarine can lay in wait at the sea bottom, or even sneak up into a fleet, it can inflict huge damage on a military enterprise. Due to this exposure, contending with submarine threat has been a major thrust in U.S. Naval strategy since WWII. Whether scanning with "active sonar" or listening with "passive sonar," anti-submarine warfare technologies depend heavily on sound.[103]

Water is fairly opaque to radio waves, and it readily absorbs and diffuses light, but it transmits sound very efficiently. Sound works so efficiently in water that a majority of marine animals have adapted to acoustic perception with the acuity that terrestrial animals have adapted to visual perception.[104] In response to the acoustic properties of the ocean, navies have developed diverse "Sonar" technologies. Sonar—the acronym for "Sound Navigation and Ranging," is really the only practical way vessels can communicate underwater, so it is used both to detect hostile submarines as well as communicate with allied submarines and other escort vessels. Sonar is also used by fleet vessels for other navigation purposes, such as depth sounding and obstacle detection. And in the event that an enemy submarine does sneak into striking range, loud sonar "jamming" signals are also used to thwart any hostile communication. With all of these various and sundry sonar systems in play, a marine warfare arena is an exceedingly noisy place; saturated with sonar sweeps, groans, whistles, pings, chirps, blasts, pops, snaps, and warbles. These deliberate noises ride on an already loud bed of engine and propeller noise and hull-coupled equipment noise, so when you see a fleet of navy ships bobbing around in a bay somewhere, the marine environment for miles around it is acoustically toxic.

Transmitting air-borne sound for over a mile takes considerable energy, but sound can easily travel for miles—even hundreds of miles in water. One communication system called "Surveillance Towed Array Sonar System/Low Frequency Active" (SURTASS/LFA in military alphabet soup) is designed to transmit sound over thousands of miles of open sea from a single platform with a source noise level of 215 dB.[105] It is considered a communication device, but due to marine animals' dependence on sound perception, it may function more like a weapon as far as sea life concerned; harassing, or even maiming animals at close range (within a few hundred meters) and otherwise acting as a medium-grade annoyance for all animals within a thousand mile radius.

While SURTASS/LFA was not designed as an antipersonnel weapon, the impact of this noise level on humans can be severe. Human ears are poorly adapted for underwater hearing and may endure high level underwater noise, but the human body couples water-borne acoustic energy very effectively, so the effects of this noise on the body are quite dramatic: U.S. Navy tests exposing human divers to 160 dB(re: 1 µPa) low frequency sound (an energy equivalent to 99 dB re 20 µPa in air) induced seizures and other long term physiological effects.[106] The military mind can take this information into all sorts of unspeakable places, and you can be assured that it has.

Sound and Healing

The words used for sound and the words used for states of health all weave through the etymologies and dictionaries together. And while "sound health" and "sound hearing" arise from different origins it is not a coincidence that sound and healing remain companionable in western languages. Sounds are used to read the health of a body when physicians palpate—tap and listen to a patient's body to diagnose their state of health. They are listening for a "sound body."

Sound is used to heal unsound bodies; from the calming effects of a familiar song on an anxious child, to the transformational power of the shaman's drum, enticing a wandering spirit away from an afflicted soul.

Music is an irrefutable panacea to bring the spirit back into the body: The mother's gentle song; the incisive rhythms of the griot; the focused oration of the cantor; and the disembodied vocal mannerisms of the shaman; these sounds envelope the listener, unifying them with their surroundings. The beat of the drum, the pulse of the heart, the flow of time: Music, like blood coursing through human experience, serves as a wayfinder for spirit and a vehicle for organic transformation.

The first healing instrument—the human voice—teaches us that sound can heal. We are calmed by the song of our Mother and assured by the voice of our Father. The power that familiar voices have in bonding us to our kin and community is probably the most persuasive argument that sound can make us whole—and heal us.

Humans have a broad range of singing and speaking voices, which we consciously use to communicate and express ourselves. But we are also capable of a vast panoply of other vocal sounds—some strange, mysterious, and unearthly. These other voices can affect us in strange, mysterious, and unearthly ways.

The power of the voice and the incantation of sacred healing words weave strongly in many sacred traditions. Sacred words and phrases, or "Mantras" are incanted to bring about results—to heal, to locate, even to destroy. This Sanskrit word is a conjunction of *man*—"intelligence" or "feeling," and *tram* meaning the protecting power—or "wings" to both shelter and give flight to the sound.

> Lama Govinda speaks of the "mantra as primordial sound and as archetypal word symbol." Mantra formulas are "pre-linguistic." They are "primordial sounds which express feelings but not concepts, emotions rather than ideas."[107]

The most ubiquitous of mantras is the sound "OM" which is chanted with a depth to encompass the extents of the creation by Buddhist monks. The *basso-profundo* chanting of Tibetan monks sung well below their speaking vocal range with an open throat; setting their entire surroundings into resonance with a sound that seems to reach into the earth's core, generating a vast range of upper harmonics which suspend around them like twinkling stars. It is here that the cellular resonance of the body takes flight. You could apply the metaphor of the vocal cords as "plucked strings," but this would only sketch out the obvious: There is vast power in the presence of this sound; a single voice sounds like many, a whole choir forms a thick harmonic landscape that you can dwell in. Practitioners believe that the sound of "OM" can bring enlightenment. Take a moment and try it out: Close your eyes and take a deep breath. Sing your exhalation into the round vowel "Oh." As you sing, slowly close your lips into the closed "mmm" sound—humming to completion. After a few times repeating this I definitely feel a pool of calm that centers my busy mind. It is definitely more sensational than chanting "bac" or "pliew" (unless these sounds are your personal mantra—sounds that resonate with you specifically). In Buddhist practice the sound from which the universe is constructed holds all things together. Chanters feel that these bonds can be transformed through resonance, so delight, wrath, orgasm, healing, and enlightenment are all available in their focused mantras.

The profound power of healing sounds is also found in early Christianity, spoken by Jesus Christ himself. There is informed speculation that Jesus may have studied medicine with the Egyptians. While the ancient Egyptians are known these days for their fabulous monumental architecture, in biblical times they were known for their medical arts.[108] Jesus' ability to heal by touch and sound was well established medical practice in Egypt of that time. In the gospel of Mark, Jesus was able to bring speech and hearing back to a man by grasping the man's head with his fingers in his hears and pronouncing "Eph'pha-tha"—Egyptian for "be opened, open your ears, loose your tongue."[109]

Since sound unseen can compel, command, inspire, and transform, there are few cultures and civilizations past or present that have not considered sound as a healing force. Songs are offered to supplicate deities, spirits, animals, and plants—where words and material offerings won't do. Chants and spells serve as pathways and bridges to the realms of the felt-but-unseen. Sounds of sacred instruments weave musical landscapes into which mortal bodies can enter, dis-incorporate and transform. The body, mind and spirit naturally entrain to organically derived rhythms, giving the conscious mind relief from the navigation of the will. With this music the body and spirit take flight, and the shamanic musician can take stewardship of an individual or a community, guiding them through diverse perspectives of reality; navigating through the heavens or underworlds to encounter the iconic and archetypal forces that inform their understandings and beliefs. It is the pulse of the shaman's drum that heals, bonding the one to the many, distracting the self from the limitations of subjective reason, unifying the bodies of the individuals into the body of the community.

Shamanic cultures are by nature invested in a sensate world. These practices remain in areas where nature is clearly the dominant force and the distractions of western technology are ineffective, rattletrap, or scarce. Surrounded by the capricious and irrefutable forces of nature, the Shaman—a naturalist of a kind, sits at the perimeter of society mediating the boundaries between the patterns of social life and the mysteries of the plant and animal world.[110] Sound is the vehicle; the Shaman's drum is often referred to as a horse that transports the Shaman between the worlds.[111] Rattles serve as a gateway—like a beaded curtain through which the healer and his patient travel; into a place where altered vocalizations speak, mutter, and growl the voices of demons and spirits. Sharp bone whistles pierce the realms of drones set down to harmonize the travelers; horns blare out to shock the perimeters—all to the steady pulse of the drum. The patient becomes saturated in a sonic fabric; heartbeat, breathing, even brainwaves entrained to the sounds of the Shaman.[112] From this altered state the Shaman can then guide the patient back, "reassembled" as a whole and well person. This is not "music." They are not composed pieces one would hum as an idyll to while away the day. These are acoustical tapestries bringing the gravity of the occasion together with a deep practiced tradition.

It is in these same societies that dancing and music are often much more than just party and social activities. Rites and ceremonies keep the community blood pulsing, bringing people together, engaged in their important roles in society.[113] A dance may be a celebration and at the same time be an important solemn event where specific tasks are accomplished: Youths are welcomed into adulthood; gratitude expressed for abundance; the arbiters of weather and the seasons are assured that the community still needs their mercy. In this context, the Dance brings all stakeholders together, making the many into one.[114]

In many African traditions, the Dancers are the "lightening rods" to Spirit. They invite the Spirits to inhabit their bodies and work their magic and medicine. The Musicians act as a conduit to that spirit power, grounding the supernatural energies flowing through the Dancers. It is by this manner that the music is "played" by the deities through the Dancers; manifesting the "spirit energies" into sound, driving the Musicians to connect the intangible spirit realms to the tangible world. The bodies of the dancers and the bodies of the instruments are conductors—alike in their capacity to channel this energy. The Musicians hold things in place. In the end, the Dancers and Musicians are grateful to each other for their respective roles in their community healing. This Dance is not a performance. Unlike western dance music, the musicians are not playing "for" the dancers; the dancers are not dancing "to" the music. Rather the resonance of sound and dancing harmonizes the community. This engagement is like food—nourishment for the collective bodies and souls. Without it the community spirit will atrophy and starve.

O.K., so you might say that "these sounds do serve to unify the sensibilities, focus the intentions, and enrich the emotions of the community. In the context that all illness has a somatic aspect, these healing practices would naturally invigorate the health of all participants. But this is not brain surgery; these sounds can't mend broken bones."

It is true that wild spirit dancing would probably be stressful on the mending of bones, but there are sounds that do mend body and bones. It has been suggested that the purring of cats may be just that sound. Cats of course purr when being affectionate, but they also purr when giving birth to kittens and mending from physical trauma.[115] This conjecture is supported by the fact that broken felid bones take significantly less time to heal than broken dog bones, and that low frequency vibrations in the range of cat purrs are used to heal complex fractures in humans.[116] Purring and healing is not specific to felids; some birds, notably puffins and storm petrels also purr to their eggs, enabling them to hatch.[117] This opens up the inquiry as to whether animals other than humans use sound as a healing energy, and if the cooing and muttering of animal mothers to their offspring serves as more than just bonding vocalizations. It fuels the speculation that in lieu of patent medicines and surgical instruments, some animals may use acoustical methods for their own health care. Given the synergistic effects of environment on health, this might be hard to measure, but interspecies sound interactions with humans may provide some clues.

There is certainly enough anecdotal evidence that dolphins produce healing effects on people, although typically not in the realms of bone and tissue repair. Dolphins vocalize in an ultrasonic range that penetrates soft tissue. They can use this sound tool to discriminate soft tissue in their prey, but it also allows them to see into the bodies of humans.[118] The accounts of captive dolphin-human interactions are rife with dolphins identifying early-stage pregnancies in women, tumor growth, and internal prosthetics such as heart valves and bone pins.[119] Dolphins also vocalize down into the higher end of human auditory perception, and the sensation of their voices can be quite thrilling, because when you are in the water with them, you don't necessarily hear this through your ears, rather you hear it through your body.

I have had the good fortune of swimming with dolphins in Hawai'i and found that their audible frequency sweeps stimulated my spine as if my vertebrae were sympathetic xylophone keys being played by their voices. I also found that these encounters—in a trusting "play field" with these amazing creatures, produced a psychological effect akin to deep meditation.

The positive neurological effects of "dolphin therapy" has spawned many "dolphin healing centers" and research into the healing effects of dolphin–human interaction on neurological disorders: Autism, Down's syndrome, cerebral palsy, epilepsy[120] and communication disabilities.[121] And while just the thrill of interacting with a magnificent wild creature would raise anybody's spirits, the effects on brainwaves have actually induced a significant decrease in brainwave frequency—akin to the effects of self hypnosis or chemical sedation[122]—benefits that can be traced to the consequences of the sound produced by the dolphins.[123]

These effects may seem subtle to the allopathic healers who conceptualize the body in more mechanistic terms. The "holistic" sound-healing models hinge on the assumption that psychological states and emotional health affect the self-healing abilities of the body. But even to those who dwell exclusively in the realm of Newtonian physics, sound nonetheless does have a role in medicine. It its most basic form, any medical doctor will listen to their patient with a stethoscope, and palpate the patient's body to check for inflammation and fluid-saturated organs. Ultrasound, or high frequency acoustical energy is commonly used for internal imaging, and it is occasionally used in the form of microwave heating to warm up organs and joints—stimulating circulation. High energy ultrasound is used for surgery to break up gallstones, kidney stones, and cataracts, and very high frequency sound is even used for brain surgery.[124]

Of course this latter isn't really sound, per-se, it is just acoustical energy. It's not resonating or aligning anything on a "meta" level; it's just setting cells in motion through physical resonance; burning holes and cauterizing tissues—allopathic applications of a subtle energy in not-so-subtle ways.

We can thank the dramatic successes of western medicine to its framing the body in physical, allopathic terms; where the physician takes control of the patient's body and almost forces healing. Of course it works, and works well. But the ancient and long standing successes of healing music and sounds have been pushed aside. In its place is a more systematic application of patent medicines and a modern form of shamanism where the healers are not bridging the boundaries between the community and the mysteries of nature, rather they bridge the boundaries between the individual body and the mysterious language of science.

But "allopathic medicine" and "holistic healing" and are not necessarily mutually exclusive concepts, rather they represent two healing modalities that arise from complimentary modes of human consciousness; a Cartesian/Newtonian framework where causes and effects occur in measurable dimensions of time and space, and a "transpersonal" experience framed by metaphor and perception.[125] In the former, the forces of nature interact in predictable and repeatable ways. In the latter, associations of elements, actions and consequences are more fluid. It is in the transpersonal realm where the experience of the body extends beyond its physical form and where sound at various frequencies can "resonate" the cells, "energy centers" or "chakras" of the body.[126] Sound readily bridges between these distinct realms of consciousness where quantifiable physical qualities unfold into imaginal landscapes over measurable time. There is enough overlap between measurable physical evidence and transpersonal constructs that the perceived effects are more than just metaphors. It is here where "hospital quiet" is promoted and the structured music of Mozart and Bach is played in convalescent wards to help patients heal.

Mountains of literature linking sound and healing date back through all written history. Even in our times the sheer volume of this literature is a "close second" to the volume of texts on music criticism—indicating that despite the desire to rationalize the conditions of the body into easily solvable physical relationships, other factors still prevail. Perhaps the pivotal question of this discussion is whether our health is strictly physiological/biochemical, or whether health states—even symptoms quantifiable in strict biological terms—are somehow connected to the larger fabric of our sense of where we are. If the latter statement is true then western allopathic medicine may be selling short on the healing possibilities of sound.

Sound and Intention

Sound is the physical signature of our dynamic surroundings. Things that move produce it; things that don't move nonetheless impinge on it. Our place in the physical world is a continuous engagement with the sounds in which we dwell and amongst the sounds we ourselves create—a condition that exists even in the bodies of those who are auditorily deaf. While we are subconsciously affected by the sounds of our surroundings, we are also consciously (or unconsciously) mediating our placement in our surroundings with our own sounds. It is a dominant tool we use to get things accomplished, whether it is a simple sigh or an involved oration; a banging on the table, or the playing of a piano concerto.

As tool makers we have crafted and refined this tool. We have taken the feedback we get from our own sounds and used it to build complex soundscapes which we inhabit. It is through this that we have created an active perceptual world; the recognition of our surroundings, the responses to our conditions, the expression of our self, and the setting of our boundaries. We aspire to affect some control over that which is within our perimeters by using sound in seductive, encouraging, forceful, or subversive ways. We delight in the engagement, crafting sound into music, communication, and medicine.

This "Song of Creation" has many singers; it is infinitely dynamic and as large as the cosmos. There is some incomprehensible order to it all—some unifying factor we call the Universe.

Singing is a manner in which we can make the element of air manifest in our bodies and out into our surroundings. Our voice takes this fundamental building block of life and helps us feel its invisible form. It is no surprise that some of the most moving human transformations are wrought from the materials of song and voice. It is no surprise that the words "inspire" and "spirit" are all born out of the wind.

What is this thing called "Sound?" **3**

> *"As our problem is to study the laws and sensations of hearing, our first business will be to examine how many kinds of sensations the ear can generate, and what differences in the external means of excitement or sound, correspond to these differences of sensation."*
>
> (Herman Helmholtz, *"On the Sensations of Tone,"* 1862)[1]

In our first two chapters we examined how we are affected by the sounds in our surroundings, and how we affect our surroundings with sound. The bridge in this transaction is the phenomena of sound itself; how it is created, how we receive it, and ultimately what happens to it once it impinges on our bodies and thus our minds.

Unlike smell, texture, or taste, sound is not a property of something; and while we can see an object that hasn't moved in a million years, for sound to exist, something has to happen. For us to hear it, quite a few things need to happen. As we examine these things, it is important to make the distinction between "sound"—what we hear, and "acoustical energy"—the physical phenomena that we perceive.

For the sake of this clarification, "acoustical energy" is considered the physical displacement within matter that is a result of a mechanical action imposed on that matter. The matter does not need to be something inhabited, like air or water that we inhabit, or the wood that termites inhabit; it can be any matter—from lead to ether.

Acoustical energy needs matter to exist. It does not pass through a vacuum[2] because there is no matter in a vacuum. If a brick is thrown in a vacuum you would not hear any sound when it hits its target. Nonetheless there would be acoustical energy within the brick once it strikes because the brick is matter; but it would stop at the boundaries of the brick without transferring this energy into the surrounding vacuum. Acoustical energy is a product of mechanical energy transmitting through matter. If the matter is something we are submerged in, like air, water, or molasses, we can sense or feel the energy through our ears and bodies. This is what we call "hearing." "Sound" is what we hear.

M. Stocker, *Hear Where We Are: Sound, Ecology, and Sense of Place,*
DOI 10.1007/978-1-4614-7285-8_3, © Michael Stocker 2013

Most animals can perceive acoustical energy through their bodies by way of "proprioceptors" and "mechanoreceptors"—sense organs that respond to body position and mechanical stimulus, respectively. Many animals, including humans, also have phonoreceptor organs—ears in our case—that are finely tuned to receive acoustical energy and transform it into neural impulses. In a broad sense, these phonoreceptors are the organs that translate acoustical energy into sound.

Understanding the distinction between these two independent phenomena—"acoustical energy transmission" and "sound perception," has long contributed to the confusion about how we think of, or imagine sound. "Acoustical energy" can behave in ways that are hard to perceive, and we can perceive sound in ways that seem inconsistent with our understanding of the material world. Our increasingly refined grasp of physics and our broadening concepts of perception are bridging the gaps in our understanding to a degree, but I find that most people's innate comprehension of how sound works has not progressed much since the eighteenth century—perhaps earlier. Some of these notions seem quaint to an informed reader, but they remain persistent to this day.

For example, it seems that there is still a prevailing idea that sound is an independent body that moves around in shapes, like an invisible water balloon, or that it is contained in some defined envelope. This leads to a common belief that hanging obstacles in the way of this "sound mass" will block it, or that there are surface treatments like "sponge paint" that will absorb sound and contain it in some manner. This echoes some of the earliest recorded acoustical science by "atomists" who explained sound as a "molding of air by the sound-imparting body" into forms shaped like the original body, or sent out as particles or fragments to find residence in the ear (and eventually the liver!) of the beholder.[3]

There is also a common belief that higher frequency sounds travel faster than lower frequencies—an idea that was first described by mathematician Archytas in the fourth century B.C.E.[4] Fortunately Archytas was wrong, because if sound velocities were dependent on frequency, sound frequencies would "time smear" with distance and we would be limited in how far away we could hear a coherent sound. Nonetheless there are some remnants of the Platonic idea that various "sound bodies" travel at different speeds or even that "sound bodies" can penetrate the body and into the brain or other organs.[5]

I haven't directly encountered the idea that sound has a will—at least in the "romantic toil" terms expressed by nineteenth century architect Alexander Saeltzer:

> ...not knowing which way to turn; poison on all sides, ever anxious to do its duty, full of natural vitality...becomes disheartened, leaves first its battlefield...and at last, exhausted, looks up as high as possible to attain rest in the strata of warmer more flexible air, air more congenial to its nature, which is always found at the highest point.[6]

...though I have encountered 20th century versions of the same sensibilities where sound mixing engineers in rock concert stadiums talked about correcting a bad acoustical problem by trapping or "blasting the sound against the back wall" of a concert hall with even louder sounds to prevent rear-wall reflections from coming back into the audience soundfield. (These guys were "sound professionals" mind you.)

Hopefully we can dispel some of these more erroneous notions by first examining the behavior of acoustical energy, and then exploring how living organisms perceive and exploit the physical characteristics of sound.

The Mechanics of Sound

As indicated above, sound is a perceptual product of acoustical energy impinging on our body and hearing organs. The sounds we perceive are affected in two realms; the spatial realm within which we hear sound, and the physical realm of the things that form that spatial realm. The spatial realm includes the shape, conditions and properties of our surroundings—the open field, the surrounding trees, the concrete canyons, the watery depths, the walls, floors and ceilings, the ground below and the open sky above—in effect the complete environment within which audible acoustical energy is at play.

The physical realm includes the acoustical properties of the things that form these surroundings and how they reflect, transmit, and absorb acoustical energy. The realm of physical properties may seem academic at the moment, but these properties can profoundly affect the qualities of the sound we hear. Material properties of our surroundings are in fact so fundamental to what we hear that they are a good starting point for our inquiry into sound perception—and the many different ways humans and other animals hear.

Acoustical energy, like light and heat, is a product of some work being done. But unlike light and heat, the acoustical energy that concerns us is not really a product of molecular motion, rather is a product of the motion of larger things. It is a consequence of bodies in motion. In most instances it is considered a by-product of movement—almost like one of the "inefficiencies" of motion. For instance, if I slide a metal chair across a wood floor I will hear a noise caused by the friction of the chair legs against the floor. This friction deflects both the legs of the chair and the floor, setting both the legs and the floor in motion. In this case, the legs and the floor material want to remain in their original form, so once deflected, they try to return to their original form. (In the language of physics, this is called the "elastic" properties of the floor and the chair legs.[7])

Depending on the flexibility and compliance of the metal legs and wood floor, their returns will overshoot a bit, and then bounce back and forth again setting up a vibration until finally they come back to rest. This vibration exists in all dimensions of the chair and floor, so you can feel the vibrations in the chair through your hand as it drags the chair, and if the vibration is strong enough, you can feel it in the floor through your feet (that is if you are focusing on the tactile sensations of the floor and chair). So this simple act of dragging a chair has set up a complex set of vibrations that are dependent on some of the physical properties of the items in play. These vibrations expand throughout the chair and floor as a by-product of the act of moving the chair.

Where the surfaces of these vibrating bodies come in contact with the surrounding air in the room, they impart their motion into the air, which is also subject to the same laws of elasticity (actually called "compressibility" for gases and liquids).[8] The surrounding air is set into motion, expanding and contracting in all directions, radiating away from all of the originating surfaces.

The phrase "expanding and contracting" is important to note here, because acoustical energy does not carry matter with it, rather it is a successive and sequential transformation of the transmission medium. Acoustical energy traveling in the wood floor (or in the air) does not move the medium per se; rather it is a flex-oscillation of "pressure gradients" within the material due to the material's elastic properties. Depending on the density and the elasticity of the material, it flexes on the transmission axis to create areas of positive or negative pressure gradients within the material (relative to the material at rest). There is a tiny bit of internal motion as a result of these pressure gradient oscillations, which reveals an additional type of acoustical energy motion called "particle motion." If you could grab a hold of the smallest particle in any oscillating material, it would move back and forth (or perhaps in a little elliptical pattern like a Styrofoam cup floating on a lake, undulating with the waves.) So in our model we have the successive spreading out of both pressure gradient and particle motion in all of the materials at play in the floor, the chair, and by consequence, the surrounding air.

Let's put this whole chair-dragging system in "pause" for a moment so we can assess where we are—because there are quite a few things happening here.

For starters, the physical energy that originated at the points of friction between the chair legs and the floor is spreading out into the floor as well as up into the legs, seat, and back of the chair. The surfaces of these energy-conveying bodies are all transmitting some of their energy into the surrounding air by force of their physical displacement, which then expands into the environment. We have previously felt the vibration through our hand in contact with the chair and our feet in contact with the floor. But at this "airborne transmission" stage we are now ready to feel the energy on the surfaces of our body, the surfaces of our ears, and most particularly on the surfaces of our ear drums.

When we put the system back in "play" again we have a whole environment—air, chair, floor, and your body all vibrating due to the acoustical energy in the system (all subject to their individual properties of elasticity). These actions all take finite and disparate amounts of time to unfold and spread out, so we will hear the scrape of the leg contact with the floor before we hear the sound of the same scrape radiating off of the back of the chair, which we hear before the coincident vibration as it radiates off of the surrounding floor. We will also feel the vibration in our hand even before the sound of the scrape reaches our ears (because it takes more time for the vibrations radiating off of the chair to reach our ears than it takes the vibrations in the chair to reach our hand).

All of these perceptual channels make for a very complex sound signature in the time domain—with coincident vibrations radiating off of diverse surfaces, from their respective displaced locations, each at their own time. And this is only the first incident sound that we perceive—the sound that is directly transmitted from the primary vibrating bodies.

The acoustical energy in the air is also imparted into the other elements of the room (the walls, other chairs, the windows, the curtains, etc.) where it is reflected, absorbed, or reradiated back into the room. Acoustical energy in the floor is also mechanically and acoustically conferred to the other elements in the room—the tables, the carpets, and the walls. These in turn will reflect, resonate with, or absorb the acoustical energy. All of these transactions will temper the sound field in the room and modify the acoustical energy set in motion by dragging the chair across the floor.

In some respect the whole scene is a bit like a bowl of jello-mold salad all set to quivering with a little bump on the side of the bowl, everything moving every-which-way for a few moments; grapes and cut cherries adding inertia to the vibration, the bed of lettuce holding the bottom steady, and the whip cream riding on top—except that we are also cast into this jello and taking part in the ride. With a little dragging of the chair we are all of a sudden suspended in a complex vibrating field of the action.

One of the dominant properties that we are examining here is elasticity—the tendency of a deflected body or compressed volume to wiggle back into its original state. Each element in this model has a unique elastic characteristic; from the metal chair legs to the air, to the walls and curtains. Each item's elastic property or "modulus of elasticity" plays into the transmission of acoustic energy throughout the system—tempering and forming the sounds we perceive—which we can do with a high degree of accuracy, allowing us to concisely identify an environment by its characteristic sounds.

In this narrative our imaginal environment includes a metal chair and a wooden floor in a "typical" room. If the floor was thick gooey mud or the chair was a block of solid stone, or the room was draped in elk hides, the sound would be entirely different—all depending on the elastic properties of the items and surroundings in play.

Elasticity accounts for how well something transmits acoustical energy. The more elastic a substance, the better it transmits the energy. Thus wood and metal are good transmitters, lead and bananas are not. Air is a mediocre transmitter, as it's "modulus of elasticity" falls somewhere between wood and bananas.

Air is also a mediocre transmitter of acoustic energy because of its low relative density. Air is much less dense than wood or metal so it doesn't transfer or "couple" acoustical energy as well as these denser materials do. Coupling efficiency—how well a material conducts acoustical energy, equates to how much work can get done with a given amount of acoustical energy in the system. One measure of coupling efficiency equates to "sound intensity." You can get a sense for this if you reach your hand out away from you and lightly scratch a nearby surface (such as a wall or table top) with your fingernail while listening to the consequent sound. If you then press your ear against the surface and listen to your scratching though the surface material, you will notice a significant increase in the intensity of the sound. When you lift your head away from the surface into the air again, the intensity of the scratching sound goes way down. There are a few factors which account for the differences in coupling efficiency between air and the table or wall surface, but density and the play of elasticity are high on the list. For a given amount of energy put into the system (scratching with your fingernail) the intensity of the sound transmission through a denser, more elastic material is greater than it is in air.

This discussion so far has expanded upon the characteristics of acoustical energy transmission through matter, influenced by the physical properties of density and elasticity. These properties are expressed in specific dimensions that can be explored in any good physics text. For our purposes just having a sense of these dimensions is adequate, but there are two quantifiable dimensions affected by these properties that will be useful to us in this discussion: Wavelength and Velocity.

Wavelength is the distance between the expanded and contracted areas of acoustical energy in a material, and velocity is how fast these "waves" move through the material.

As I previously mentioned, the speed of sound is not dependent on frequency. High frequencies and low frequencies all travel at the same velocity in a given material. But wavelength is dependent on frequency, with the relationship of "the higher the frequency the shorter the wavelength."

The "pool of water"[9] analogy is commonly used to illustrate this. In this example, if you drop a stone into a pool of water, waves will radiate away from the center of the splash. The distance between the wave crests is called the "wavelength." If you drop a pebble into the pool, it will generate ripples with a wavelength of only an inch or two. If you drop a large river stone into the pool it will generate larger waves with a wavelength of perhaps a few hand-spans.

These waves will radiate away from the splash at the same velocity, so if you poke a stick into the mud near your rock-drop area, the crests will pass the stick at the same velocity, but the short wavelengths will pass the stick more frequently (higher frequency) than the longer waves. This relationship is tidily expressed in the following:

$$\text{Wavelength} = \frac{\text{Speed of sound}}{\text{Frequency}}$$

or:

$$\text{Frequency} = \frac{\text{Speed of sound}}{\text{Wavelength}}$$

From this we see an immutable interdependence between frequency, wavelength and transmission velocity.

Transmission velocity on the other hand, is not the same in all materials. This is due to their differing properties of elasticity and density. Expressed mathematically this relationship is[10]:

$$\text{Speed of sound} = \sqrt{\frac{\text{Elasticity}}{\text{Density}}}$$

Thus the speed of sound varies depending on the material, and that at a given frequency the wavelength will be different in differing materials due to the differences in the speed of sound in the various materials.

These basic relationships have been expounded upon and detailed by physicists, but this simple form is suitable for armchair mathematicians. What this implies is that the transmission speed (and thus the wavelength) of acoustical energy is different in every material, and is dependent on the properties of that material. To get an idea of the range:

Velocity of sound in various materials[11]		
Material	Velocity in ft/s	1 kHz wavelength in ft
Air at sea level	1,120	1.12
Fresh water	4,820	4.82
Sea water	4,939	4.94
Oak	12,872	12.87
Glass	16,733	16.73
Lead	6,950	6.95
Iron	19,147	19.15
Beryllium	41,480	41.48
Cork	1,609	1.61

With the exception of the hearing environments of air and water, velocity differences of various substances don't really come to bear on human needs for acoustical information, but it does affect various animals in terms of how the velocity of sound affects wavelengths in their subject environments.

By example: For a given frequency, the wavelength in seawater is about five times longer than in air so sea animals have adapted to these long wavelengths in the ways they use sound. In a similar manner, the wavelength of a given frequency in soil is quantifiably different than its wavelength in air (depending on composition). This feature plays into how moles locate their prey and elephants communicate through the earth over long distances[12] (features that we will examine in the next chapter).

There is another dimension of the wavelength/frequency characteristic which also plays into our perception of sound which has to do with the "net energy" in an acoustically vibrating system. Going back to our "dropping stones in water" illustration we noticed that dropping larger stones in the water produced lower frequency, longer wavelength energy. The water displacement of the large stones was greater and produced bigger waves in the water than the pebbles did. This is consistent with our experience of low frequency as in indicator of large things happening; larger things produce more energy, larger wavelengths transmit more acoustical energy. With this cognitive resonance we are more likely summoned to protective alert when we hear low frequency rumbling—even if it is a quiet rumbling, because low frequency noise is a harbinger of big things coming around—earthquakes, thunder storms, angry mammoths, or stampedes of bison. These are all big things that we would want to prepare for. Even while we are put on alert when we hear the high pitched buzz of a mosquito, it is only an "annoyance alert," not a "danger alert." On the other hand, if mosquitoes rumbled like freight trains they would trigger our "danger alert" and put our summer evening picnics in a sorry state. Fortunately mosquitoes only buzz, and we only hear them when they are within our sphere of mutual influence.

This brings up another frequency/wavelength dependent quality of sound; unlike the close-range, high pitched whine of a mosquito, we can hear low frequency, long wavelength sound at considerably longer distances. Low frequency sound has more "reach." This is a nice feature because it gives us advance warning of big things coming our way. The long reach of low frequencies is a result of two characteristics, one a function of the higher energy density typically found in low frequency sound, the other is a property of the environment and how it absorbs short wavelength, high frequency sounds and transmits long wavelength, low frequency sounds. High frequency wavelengths interact with, and are absorbed by the many small things in our surroundings; blades of grass, leaves, clothing, even air itself. On the other hand there are few things in our environment large and supple enough to absorb long wavelength energy, so it travels pretty far before it dissipates.

You can really get a sense of this when you are subjected to some "car cruiser" with a loud sound system in his "sled." Initially at a distance you will only feel the deep bass "thuds." As it approaches, the higher "whacks" of the drums will start emerging. You may not hear the melodic instruments until the car is close enough to throw rocks at, and won't hear the high swishes of cymbals or the pitches of lead instruments until the car pulls right up next to you. This "filtering with distance" effect is a product of how the environment absorbs acoustical energy; because the small features that absorb high frequencies are more prolific in our environments, you will hear less high frequency sound as you get further away from the source.

The qualities of "sound absorptivity" (and the complimentary "sound reflectivity") express how materials affect their contiguous acoustical environment. These are environmental qualities that concern most architectural acousticians because they affect the reverberation characteristics of acoustical spaces. "Reverberation" is an expression of the length of time that acoustical energy dwells in a space before it is absorbed, and has a bearing on how speech, music, and sound-comfort are enhanced or compromised by the qualities of an enclosed space.

As a rule, hard, flat, high density surfaces reflect sound, and fluffy, thick, low density materials absorb sound. A basic understanding of these characteristics is suitable for most situations, but the discussion does not end here. There are gradations of absorptivity (and reflectivity) which are usually expressed in terms of "absorption coefficients." These gradations are not linear across all frequencies; rather the absorption coefficients of various materials and structures are "frequency dependent." Some materials absorb short wavelength, high frequencies better than they absorb long wavelength, lower frequencies, and some materials absorb only certain frequency bands while reflecting others. This all has to do with the geometry of sound—the wavelength dimensions, and how these dimensions interact with the geometry of the particular absorber/reflector.

A good starting place for understanding absorptivity is with the unit of measure used to quantify it. This unit is called the "sabin," in honor of Wallace Clement Sabine (a methodical Harvard professor who by happenstance became the godfather of modern acoustics).[13] The "sabin" is a unit of perfect acoustical absorptivity, and is defined as a "one square foot open window."[14] The open window is a good benchmark from an architectural acoustics standpoint as it describes a planar surface area that architects associate with walls and wall treatments. (From an environmental acoustics standpoint though, the sabin loses some of its utility, especially once three dimensional sound absorbing forms are brought into play, interacting with the dimensions of wavelength.)

Usually when we speak about absorption we are talking about sponges and towels and how well they collect liquid and hang on to it, but sound absorption doesn't quite work this way. Sound absorption is actually an energy conversion process and refers to how well a material converts acoustical energy into heat. It seems odd to think of it this way unless you are a physicist, but when acoustical energy impinges on a material or structure, it sets it into motion. The mechanical frictions induced as the material flexes will generate heat. We can't hear heat, so it is through the flexing of the material that sound is absorbed and a miniscule amount of heat is generated and dissipated into the material. The fact that you can't feel this heat and can only measure it with the most sensitive lab instruments attests to the delicacy of airborne sound and the efficacy of our senses to perceive it.

Wavelength, frequency, velocity, absorptivity, and reflectivity come into play

This discussion about the conversion of sound into heat brings us closer to the physical dimensions of sound—and how sound absorptivity is frequency dependent. It is here where the actual wavelengths come back into play.

Because of their wavelengths, specific frequencies interact with specific dimensions of materials and structures. Thin and delicate materials like silk and satin interact with short wavelength, high frequency sound; bigger fluffy materials like cotton batting or fiberglass insulation interact with longer wavelength, lower frequencies. This all works because the physical dimensions of a material allow it to resonate at specific acoustical wavelengths.

To get a good feel for how material thickness and loft affects sound absorption, grab a big beach towel and stand about a foot away from a sound reflective wall, facing the hard surface. If you speak directly at the wall you can hear your voice reflecting back at you. Try using a series of vocables like "ta ta ta ta" or "ti ti ti" because these sounds include both high frequency plosive and lower frequency formative sounds in them. As you belt out this stream of nonsense, lift the towel up against the wall between you and the wall; you will notice that the intensity of the higher frequency "T" sounds attenuates or "rolls off." If you double the towel over and do the same thing, you will notice an even greater effect. Double it over again and the effect will likewise increase. As the towel layers get thicker they absorb proportionately longer wavelengths. At the four layer point most of the sound you hear is through your own body and mouth, and not sound reflected back at you from the wall. In mechanical terms you have created a low-pass, high-cut filter like the treble control on your car stereo.

This works well in attenuating higher frequencies, but longer wavelengths require larger absorptive dimensions, so there are practical limitations as the

Table 3.1 Frequencies and wavelengths in air

Frequency in Hz[15]	Wavelength in ft–in.	Description
20	56′	Lowest human pitch discrimination
27	41½′	Lowest note on an 88 key piano
80	~14′	Lowest range of a human bass singing voice
110	~10′	Average man's speech fundamental
220	~5′	Average woman's speech fundamental
261.63	4¼′	Middle "C," Lowest pitch on an orchestra flute
300	~3½′	Average child's speech fundamental
440	2′–6½″	International "A" pitch
1,146	~1′	Highest range of a soprano singing voice
270–3,500	~4′ to ~4″	Range of formative pitches in human speech
2,000	6¾″	Average distance between human ears
4,000–8,500	~3″ to 1.5″	Range of sibilance in human speech
3,136	4.25″	Highest pitch on a violin
4,186	3.25″	Highest note on an 88 key piano
6,000–18,000	~2″–3/4″	First order harmonics of a violin
17,000–20,000	7/8″–5/8″	Highest human pitch discrimination

wavelengths get longer. Table 3.1 gives the wavelengths of various airborne sounds in the range of human hearing.

You can surmise from the table that due to their wavelengths, lower frequencies are progressively more difficult to absorb than higher frequencies. In cases where low frequency sound attenuation is required (air conditioning noise control for example), various methods of isolation and attenuation are employed. This is the playground of architectural acousticians and where a preponderance of their income is generated.

The relative difficulty of absorbing low frequency sounds—graduating toward the relative ease of absorbing high frequencies accounts for the subtle auditory cues that give us an idea about how far away we are from various sound sources. As the sounds of voices (or machines or cars or animals) travel across the landscape, grass, leaves, trees and other features of the landscape progressively strip off the higher frequency components of the original sound, allowing us to gauge the relative distance of the sound source by comparing the received sound frequency profile to sounds stored in our internal sound maps.

We can do this with uncanny accuracy, so even while I am sitting in my indoor workspace I can hear my neighbor speaking outside on his phone somewhere in our shared back yard, and I can pretty much tell how far away—and even where in the yard he is speaking. My ability to locate him in this auditory/imaginary space is accentuated by the how the environment modifies the sounds of his voice.

Our shared back yard is an open lawn with few obstacles between it and my house. It is surrounded by high vegetation on two sides, with my house and his house constituting the major reflecting features of the other two sides. One of the vegetated sides is supported by a redwood fence; the other is open along a seasonal stream bed. All of these features reflect and absorb the sound in their characteristic ways, giving me a sense of what the auditory scene is.[16] If the back yard was covered with snow (a rare occurrence in coastal California) or yards and yards of parachute silk (more likely) I would hear these physical conditions by how they would affect the sound of my neighbor's voice arriving at my workspace. This is due to the different sound absorption coefficients of the vegetation, the snow, the parachute silk, and the other features at play in the yard.

The above discussion has treated acoustical energy within two realms of action: the spatial realm in which we hear it and the physical realm of the things that form the spatial realm. The characteristics of these two realms influence the qualities of the sound we hear, which are all framed within a third realm of action—the realm of time. A sound takes time to unfold, so it helps us gauge the passage of time. Because it unfolds in space, the quality of its unfolding helps us gauge the qualities of space. Sound perception framed by the unfolding of time gives us a dynamic understanding of the location and dispositions of all things within the realm of our hearing—our "auditory scene." This includes the ongoing dynamic events, such as the buzzing fly clicking against the window, the airplane overhead, and the neighbor's children playing in their yard; but it also includes the acoustical play of the wall behind me, the ceiling above, the pile of blankets in the corner—even the sound of my computer keystrokes clattering off of the screen directly in front of me. I gauge the relative location of all of these things predominantly in the time domain.

Acoustical environments are tempered by how their absorption characteristics influence their soundfield. Architectural materials are qualified by their sound absorptive or "noise reduction" characteristics expressed in terms their "Noise Reduction Coefficient" (NRC). But equally important to absorption is its reciprocal; reflection—which describes the amount of airborne sound that is bounced back into the environment from the surfaces it hits (and how much time it takes to eventually dissipate). Sound reflectivity is a characteristic of walls, floors, ceilings, caves, granite canyons, and even the leaf surfaces of a broad-leaf forest. These elements all have reflective components that affect airborne sound—echoing or refracting the sound in ways particular to each surface.

A perfect sound-reflecting surface would be infinitely large, too dense and heavy to be moved by the acoustical energy bouncing off of it and too shiny to diffuse any of the energy. If you were within a huge sphere built from this material, acoustical energy would bounce around, or "reverberate" in this sphere forever—if not for the friction of the air molecules transmitting the energy. (Eventually the air friction dissipates the acoustical energy into heat.) "Reverberation" is the term used to describe how long acoustical energy bounces around in an environment before the energy is dissipated. Long reverberation times are characteristic of caves and cathedrals, short reverberation times are characteristic of coat closets and shower stalls. A reverberant field is tempered by the sound reflective qualities of its constituent materials which have a profound bearing on our subjective experience of that environment. A deep cave or cathedral may yield a mysterious or somber reverberant sound field, a gymnasium or a subway station (with the same reverberant dwell time) may yield a chaotic and nervous reverberant sound field; the reverberant sound field in a shower stall may feel encouraging, whereas the reverberant field in an elevator may seem intimidating (even if you are alone in it).

Stereophonation and Localization: Hearing Space

Humans have two ears, one on each side of our heads. This allows us to compare two different, but relatively similar inputs and derive vast quantities of spatial information from the differences. If something is producing sound fairly close to you, its location can be determined (or "localized") by the relative amplitude difference between your ears. Both the proximity to one ear and the "acoustic shadow" of your head will help you determine which side of your head is closer to the sound. The acoustic shadow changes the sound spectrum because your head, hair, shirt and all of the other elements in the direct sound path absorb and modify the sound between one ear and the other. There is also a subtle time difference called the "inter-aural time difference" representing the difference in time it takes for sound on one side of your head to travel to the other, which your mind can discriminate. This path can be as short as 7″ and take approximately 500 μs (500/1,000,000ths of a second) to traverse.[17]

As something gets further away from you, the inter-aural time difference and the shadowing effects are less pronounced. This is because the inter-aural differences are diminished with respect to ear-to-ear versus ears-to-source ratio. So when we localize more distant sound sources other perceptual factors come into play. An important cue is that sounds further away from us are more "in the room" than from any specific location, so I can only determine the approximate location of distant sounds by how they play into my surroundings.

A big part of what accounts for the sound differences of various environments is the way sound bounces off of or is absorbed by the surroundings. If there is a soft wall behind me and a hard wall to my side, these will temper the sounds at play in the room correspondingly and give me a sense of where the original sound is coming from. The various surfaces and textures in the room modify the spectral content of the sound—absorbing various frequencies and reflecting others. When a coherent wave-front strikes and reflects off of a surface it diffracts in the various vectors coincident to its angle of incidence. These reflecting wavefronts bounce into and interact with oncoming wave fronts creating interference patterns that temper the sound by virtue of the interference. This auditory scene gives us a very complex set of sound cues which help us hear where we are.

Herman Helmholtz expresses the complexity of the interacting waves by using the "waves in water" metaphor, and then confers it to an airborne environment:

…This is best seen on the surface of the sea, viewed from a lofty cliff…We first see the great waves, advancing in far-stretching ranks from the blue distance, here and there more clearly marked out by their white foaming crests, and following one another at regular intervals toward the shore. From the shore they rebound, in different direction according to their sinuosities, and cut obliquely across the advancing waves. A passing steamboat forms its own wedge-shaped wake of waves, or a bird, darting on a fish, excites a small circular motion. The eye of the spectator is easily able to pursue each of these different trains of waves, great and small, wide and narrow, straight and curved, and observe how each of these passes over the surface, as undisturbedly as if the water over which it flits were not agitated at the same time by other motions and other forces…

We have to imagine a perfectly similar spectacle proceeding in the interior of a ballroom, for instance. Here we have a number of musical instruments in action, speaking men and women, rustling garments, gliding feet, clinking glasses, and so on. All these causes give rise to systems of waves, which dart through the mass of air in the room, are reflected from its walls, return, strike the opposite wall, are reflected again, and so on till they die out. We have to imagine that from the mouths of men and from the deeper musical instruments there proceed waves of from 8 to 12 ft in length, from the lips of women waves of 2–4 ft in length, from the rustling of the dresses a fine small crumple of wave, and so on; in short, a tumbled entanglement of the most different kinds of motion, complicated beyond conception.[18]

You can examine the visual and auditory ideas of this description if you fill a crystal wine goblet with water (actually a dark Merlot provides a good visual contrast), set it on a stable surface and do the "glass harmonica" trick to produce a tone by rubbing a wet finger around the rim. If you look across the surface of the beverage you will see some beautiful patterns which are caused by the oscillations of the sides of the glass flexing to the resonance of the filled glass. These fine little waves don't just radiate in straight lines or concentric circles from the edge or center, rather they sweep around and form intricate patterns of interference. From the perspective of your ear you are hearing the sound of the tone produced directly under your finger, but you are also hearing the sound radiating off of the sides of the glass, the sound resonating within the glass and the sound reflecting off of the beverage surface. This tone is reaching you from all of these sources, all spatially coupled in a complex time domain field, producing a miniature model of what occurs in the chair dragging exercise that introduced this section.

What is tidy about this experiment is that the sound is a relatively pure tone, and not a concatenation of a bunch of noises. It is also occurring in close proximity to your ears so you can really hear the effects. You may also notice the ethereal quality of the sound. This is a result of the delicate interference of the wavefronts that your ears and mind are attempting to integrate into a coherent localized sound source. The oscillations of the goblet produced under your finger along with the reciprocal oscillations of the glass opposite your finger present different wavefronts, displaced in space by the width of the goblet, and thus displaced in time by the finite time it takes sound to travel across that distance.

If you have two wave fronts of the same frequency that are offset in time by a value that is shorter than the wavelength of that frequency, they are not considered sequential events, rather they are considered to be "out of phase." As one side of the glass produces a positive wavefront excursion, the opposite side is tending toward a negative excursion. If they are exactly opposite in phase, the positive and negative pressure gradient excursions cancel each other out and no sound is heard. But if they wobble around each other as they are doing in the glass, they cancel or accentuate depending on their relative phase. The resulting sound wobbles around as well, following this geometric interference of similar shaped wavefronts in time and space.

Fig. 3.1 Wineglass. Sinusoid waves sum "in phase," cancel "out of phase," and interfere summationally between these two extents

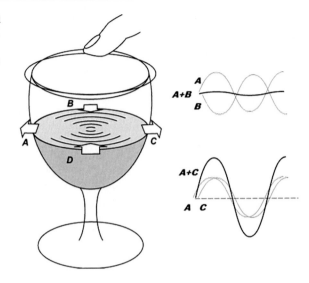

This ethereal "phase modulated" effect is very pronounced in the glass harmonica because the wave shapes are so similar and are so close in amplitude. The only quantifiable difference between these waves is their point of origin, so when these waves interfere, they do it in spades. This same time domain phenomenon occurs in pretty much all sound we hear—it's just not as obvious because most of the sounds include a multitude of frequencies, pitches, amplitudes, envelopes, and wave shapes and are thus more complex than a clear tone radiating off of a pristine wine goblet.

The pure tone of the wine goblet is unusual in nature. With very few exceptions, most other sounds we hear are a complex assembly of harmonic motions at various amplitudes and phases. It is this sound composite of all noises that helps us identify the characteristic sound of any sound-making action. We call this characteristic "timbre," which describes the way the composite sounds of an action play upon each other into a complete sound envelope. Timbre helps us differentiate the sound of our chair dragging across the floor from the sound of a clarinet or a telephone. It also helps us differentiate a clarinet played by Benny Goodman and the same instrument played by Lester Young (or by me if I could get my hands on it).[19] It is the subtle interplay of these composite sounds both in amplitude and in phase that constitute the timbre or "tone" of a played instrument, and accounts for why a Stradivarius violin is priceless and my violin isn't. Somehow in the "Strad," all of the composite violin sounds we like to hear are reconciled to the radiating body of the instrument in such a way that they issue forth magic; all of the pitches and the harmonics are in a delightful balance, and their phase interplay happens in nice ways.

It also happens that this subtle phase or time domain information in a sound helps us really localize its source. We know the sound of a clarinet because of the constituent tones that make up the whole clarinet sound. We know that the clarinet sound is coming "from over there" because of amplitude and arrival time differences between our ears. We know how far away and from what direction because of the way our surroundings reflect and absorb its sound. We know exactly where it is coming from because of the time domain or phase information imbedded in the sound that reaches our eardrums. Even with blinders on and one ear plugged, we can fairly accurately pinpoint the location of the clarinet—or most other sounds for that matter. This is a testimony to how accurately we integrate and reconcile the dynamic phase and time domain information imbedded in the sounds we hear.

We will examine how we do this in the next section, but for the moment if we head back to our chair-dragging scenario we are presented with a complex set of sounds that originate from a mechanical action. These sounds bounce and reflect around the room, time-smearing off of the walls, windows and floor, absorbing into the curtains and upholstery, and eventually dissipating into the volume of the room. This sound scene is very complex; it contains all of the elements we need to complete our imaginal participation in the action. Presented with even a monaural recording of this sound we could likely reconstruct the whole event in our imaginations. We could estimate the dimensions of the room, the size and shape of the chair, even the direction and distance that the chair was dragged. A silent video of this very same event would produce nowhere near the sense of place, belonging, or participation that the simple sound reproduction would. This speaks volumes about how we use sound to verify where we are and how we construct our maps of belonging—our "auditory scene." For us this odyssey begins the moment acoustical energy from the dragging chair touches our body. Now have a seat and we can dig into our ears a bit.

Hearing

Hearing is a vast biological enterprise which has been a source of speculation for thousands of years. The depth and breadth of the inquiry, at least in terms of physiology, has outstripped the inquiry on any of the other human senses. In the time of Aristotle philosophers attempted to describe the senses because their descriptions established the basic elements of their philosophical experience. Before cameras, microphones, tape recorders, and oscillograms, humans had to rely exclusively on their own perceptions, and on "natural philosophy" to describe their experience.

Vision, for example, was considered an illumination generated from the fire within the organs of the eye. The gaze was a collaborative act between the object and the beholder. What was beheld came into the realm of the mind.[20] Some of the arguments at the time detailed whether the eye was an "organ of water" that mirrored the scene, or an "organ of fire" that illuminated it. Aristotle refuted the conjecture that light traveled at some finite speed. These descriptions of vision served Western civilization well up into the twelfth century.

The sense of hearing was not so easily nailed down. This is likely due to the ethereal and tactile nature of sound and how the perception of sound played into the cognition of time and space. It was clearly a physical force that required the physical interaction of things, and as such it revealed the properties of these things. But it also interacted with the soul and the mind in ways that animated the material of sound. Did this mean that sound was contained in things, to be released with some interaction? Was sound an expression of a soul? If so, did bronze have a soul but not wool? Or did soul dwell in things such as harp strings and bronze shields that could be teased out through crafted proportionality?

Once this soulful force impinged on the hearing organs, it did not stop; rather it dwelled in the mind and blended with the soul of the beholder. Did this mean that sound was composed of soul? Where did sound go once you couldn't hear it any longer?

It was also believed at the time that the structure of the universe keyed into sound, harmony and "celestial music" or "Music of the Spheres." This ancient Greek belief was refined by the astronomer Kepler in the seventeenth century—that the transit of the planets produced divine music.[21] But if the planets did make music as they orbited through the heavens, why couldn't we hear it? These and many other unresolved questions posed many logistical problems for scientists and theologians alike. The interplay of Western ideas about the phenomena of sound and hearing trips through the works of many renowned Western thinkers: Archytas, Democritas, Pythagoras, Plato, Aristotle, Euclid, Augustine, Kepler, Erasmus, Descartes, Newton, Pascal—throughout and beyond—all attempted to describe this physical-ethereal interaction of sound and hearing, and reconcile it to the human perceptual reality.

None of these great minds actually succeeded, so even in our information age, when the discussion of acoustic energy has been reduced to physical concepts that come into the compass of our technical instrumentation, the actual play of sound in time and space is still complex beyond our comprehension. At best our acoustical models are approximations of what is occurring in our auditory soundfield.

If you maintain that all human perception can be reduced to neural-chemical impulses, then the incomprehensible complexity of a soundfield is echoed in the equally complex train of perception—from the reception of acoustic information, the translation into neural impulse, then into the mental artifacts of perception and understanding. Our current approximations give us a vast array of repeatable tools to begin with, but even in these times of electro-microscopic examination, many basic actions in the train of hearing are only placeholders for a future understanding. As an example of this, we are still unsure how the cochlea actually works. There is disagreement as to whether it acts as a traveling wave guide (wherein the wave-shapes of sound are condensed into a physical analog within the cochlear spiral),[22] or if it serves as a resonating structure (wherein the cochlear hair cells resonate to specific frequencies, dependent on their location along its length).[23] In light of these vague understandings, the truth is probably contained in both, and a bit of something else.

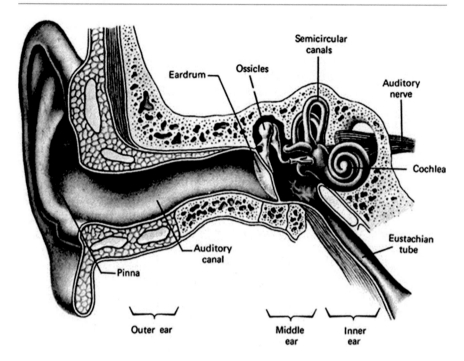

Fig. 3.2 The human ear (source: John B. Palmer, Anatomy for speech and hearing)[25]

Suffice it to say, that given the rich and deep legacy of inquiry into the sense of hearing, attempting to condense and redescribe all that is already known would be folly. At best I may be able to sketch out a simple model of how we hear, in hopes that any new ideas presented herein might add a little flavor to the soup, framed with a caveat from Albert Einstein: "All our science, measured against reality, is primitive and childlike."[24]

Human hearing organs

Hearing is a translation of physical/mechanical energy into electrochemical (or neurological) impulses. The dominant fulcrum over which this translation occurs in humans resides in the ear. From a structural standpoint, the membranes, bones, tissues and fibers of the ear reach into the skull from the outer extremis of the pinna into the inner sanctum of the cochlea. From a neurological standpoint this structure is entwined with bundles of nerves from the bones of the middle ear, the cochlea, and into the brain through the auditory cortex. Although the entire depth of this metamorphic mechanism is less than 2″ long, the translation that occurs across that short distance is profound.

The ear does not equate with the elegant simplicity of the eye—a lens that captures reflected light, adjusting for brightness and focusing the light onto photoreceptive rods and cones. The ear is much more complicated. Within it we have receptors of physical displacement, sensors of pressure gradients, phase and frequency filters, integrators of time, mechanical damping, muscles that tailor and limit the range of displacement, tissue forms that collect, split and reflect acoustic energy, cavities that mediate acoustical impedance, membranes that confer the affects of fluid motion, canals that translate meta-physical motion into bearing and balance, vents that equalize middle ear pressures to atmospheric conditions and pathways that integrate skull and bone conducted vibration. And this only outlines the physiology of the ear. Each of these physical mechanisms has neural sensors that gather, translate, mediate, and feedback neurological activity between the brain and the stimulating mechanism. And then there are the nerves that exclusively transmit acoustical-source information into the hearing centers of the brain—thus opening the gateway to perception, mapping, thought, and emotion.

Viewing the ear as a fulcrum between physical acoustics and sound perception helps me encapsulate the great complexity of the ear into a comprehensible idea. It also helps me distinguish the physical from the neurological, with the fulcrum being an assembly of bidirectional translations between the two. This model takes a "Rube Goldberg" assembly of task-specific tissues, bones, and nerves and integrates them into the holophonic entity of sound perception.

Sound perception is commonly described as a sequence of events that result in our identifying and categorizing the sounds we hear. But hearing is not really a "serial processing of incoming information," rather it is a system that processes in both directions, receiving stimulus and tailoring the sensing systems in response to the dynamics of the stimulus. By way of example: there is a muscle (*tensor tympani*) that attaches to the malleus bone of the middle ear and adjusts the flexibility and motion of the middle ear ossicles. Its action helps us accommodate for acoustical amplitude changes and protects our ears at the onset of loud, potentially damaging noise.[26] The effect of this muscle could be likened to the "wince" of our eyes when we prepare to view damaging physical impact. The neural pathways that mediate this action are complex and interconnected to a variety of nerve systems— including the cochlear, trigeminal, facial, and the superior olive systems. (You don't have to remember all of this.) The entire system presents a complex neurological environment that not only tailors physical responses of the ear to acoustical amplitude, but also adjusts the middle ear to body movement, eye closure, speaking, physical stimulation of the outer ear, and local wind velocity.[27] This indicates that the system mediating the tensor tympani muscle is tied into a larger auditory paradigm, and that a simple protective response to excessive noise is part of a larger array of information which informs us about the nature of our surroundings. Thus hearing is not just a chain reaction that serially transmits physical motion through a set of nerve sensors to be routed to the various processing centers of the brain; rather the whole system of hearing is like a neural-physical resonance between perception and the surrounding physical environment.

The motions and neural streams that confer this process are exceedingly fine. From the acoustical side of the fulcrum, vast amounts of subtle pressure-gradient activity in four dimensions (x, y, z, and time) have to be received and permeated into the sensing mechanisms of the ear. On the neural side, all of these stimuli need to be routed and sorted out to reconcile a complete and contiguous sound stream to the mind. Throughout this, the physical actions of the transfer system need to be dynamically tweaked and tailored in ways that are consistent with conditions on both sides of the fulcrum—adjusting to the external environmental conditions based on the cognitive "need to know." Not being a neurologist I can't comment on the relative magnitudes of neurological activity, but from a physical standpoint, absolutely miniscule amounts of physical motion impinged on the eardrums can trigger vast amounts of activity in the organism.

To get an idea of this, consider that the ear can detect motion in the eardrum which is about 1/100th the diameter of a hydrogen atom.[28] If this miniscule eardrum motion is caused by the close-stalking movement of a predator, almost instantaneously this miniscule motion is amplified into an evasive action of the whole body—precisely away from the source of the sound. This is a "high gain" system with a precise, delicate and fast input.

For the ear to be this responsive with the required accuracy, it needs a stable platform. So when I speak about the physical form of the ear, I like to start with the fact that middle and inner ear is protected inside the cranium and contained in an

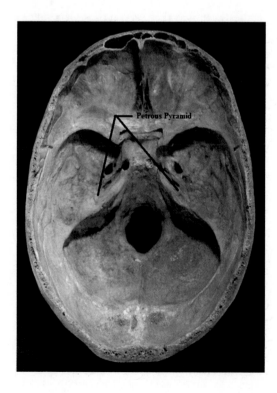

Fig. 3.3 Cranium interior illustrating the location of the petrous pyramid (photo by author)

envelope called the "petrous pyramid." Pyramids are very stable platforms, with a large broad base. The base of this pyramid is along the mandibular hinge as part of the temporal bone, sitting in the seat of the cranium.

It is "rock hard"—"petrous"—the hardest and densest bone in the body except for the teeth.[29] The purpose for its hardness doesn't seem to be a safeguard against impact damage, as it is buried and well protected within the hard skull; rather it is hard, dense and stable for precision. Like any good machinists platform, its density provides a stable, vibration and resonant free stage from which to gauge the sounds we perceive. The petrous pyramid houses the cochlea, with the mechanical systems and nerves that translate miniscule acoustical actions into neural impulses. If this platform was soft and sloppy, the accuracy of our sound perception would be no better.

Of course having a stable platform is useful, but we need to know where the platform is. It is easy to locate ourselves in our environment; we can see our hands, and gaze into the horizon. But this seat of our own location—this center of place—is concealed from all but its own workings. Fortunately it does have a clear self-reference system, because within the petrous envelope is a set of semicircular canals that gauge rotation vectors on the x, y, and z axis, and the *utricle* and *saccule* connecting the canals that provide sensitivity to linear motion and gravity—just like a gyroscope-stabilized inertial platform on an airplane. This elegant formation can be talked about at length, but this is really where a picture is worth a thousand words.

There is a vast array of morphological differences in the ears of all vertebrates. Fishes have no outer ears; turtles have outer ear membranes only; humans have stationary outer ears; bats have fabulous convoluted outer ears; mules have long movable outer ears; cats, sharks, rats, salamanders, lampreys, and lizards… each and every ear is significantly different, but all of these animals have semicircular canals.[31] To me this is the clearest physical evidence that the entire ear is an organ of location and placement, anchored in place by the semicircular canals.

The semicircular canals are filled with fluids and are lined with various hair cells that act as "strain gauges." They also contain various masses or concretions of calciferous crystals—sand and bones suspended in the fluids that serve as inertial bodies to the hair cells. Essentially accelerometers, they respond to head rotation,

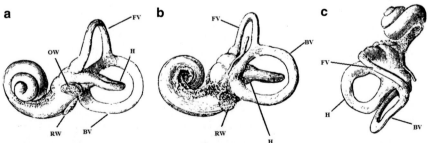

Fig. 3.4 Helmholtz-CC-canals. Labyrinth and semicircular canals: (**a**) Left labyrinth (superficial view), (**b**) right labyrinth (visceral view), (**c**) Left labyrinth (dorsal view). *OW* oval window, *RW* round window, *FV* front vertical semicircular canal, *BV* back vertical semicircular canal, *H* horizontal semicircular canal (from Helmholtz)[30]

linear motion, orientation with respect to gravity, vibrations due to seismic activity (or just plain jumping about), infrasound, and acoustic energy. There is also evidence that these little accelerometers are also able to self mediate through mechanical feedback—adjusting the damping and acceleration response of the system according to the magnitude of the motion.[32]

To get an idea of how sensitive and selective this system of orientation is, fix your eyes on a single letter on this page and move your head around, nodding and tilting, or shaking it wildly. Unless you were shaking too wildly, you were able to keep a fix on the letter. It is the localizing cues from your semicircular canals which drive the muscles that rotate your eyes so that your vision remains stabilized. Considering that the retinal fovea spans approximately 1° of arc, fixing with this precision is quite an accomplishment.[33] This degree of precision exists throughout all of our typical motions, plus a huge operating margin to provide for those ranges of motion that are not typical—like flying a high performance jet fighter or falling off a log with a bag of groceries in your arms. All of the axes' of motion are tracked through these maneuvers by way of the features in your inner ear. But the accuracy with which the fighter pilots executes turns and spins, or the speed in which you catch your balance AND catch the flying groceries would not be possible if your inner ears were not housed within the stable "petrous pyramid" platform.

The Mysterious Cochlea

Of the tubes, membranes, and channels that reside within the petrous pyramid, probably the most recognizable element is the cochlea—the little snail shell that we mostly hear with. The cochlea is unique to mammals and as I've indicated earlier in this chapter, there is no clear agreement as to how it really works. As a result it is

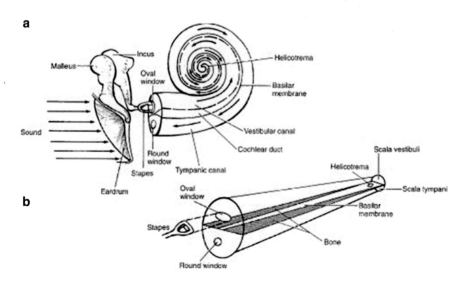

Fig. 3.5 Cochlea unrolled (Yin, Tom C.T., "Audition" 2008)[34]

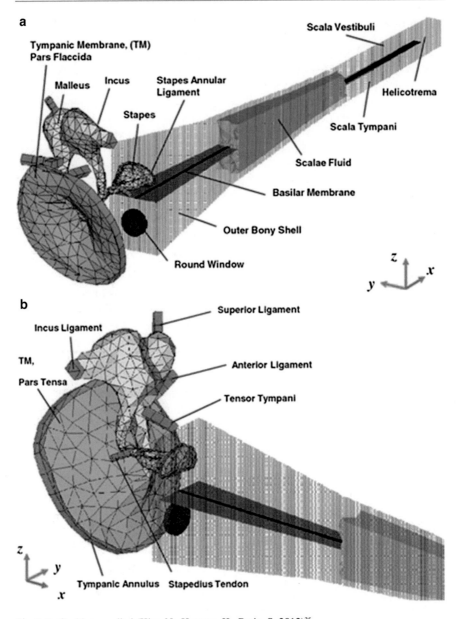

Fig. 3.6 Cochlea unrolled (Kim, N., Homma, K., Puria, S. 2012)[35]

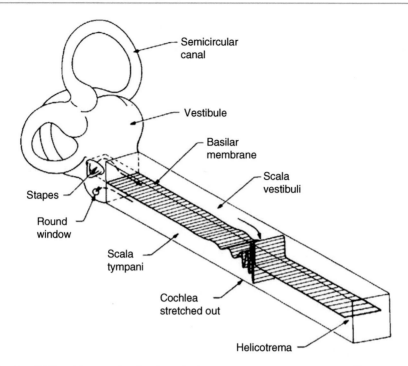

Fig. 3.7 CRC cochlea unrolled (Lewis, E. R., Leverenz, E. L., Bialek W. S. 1985)[36]

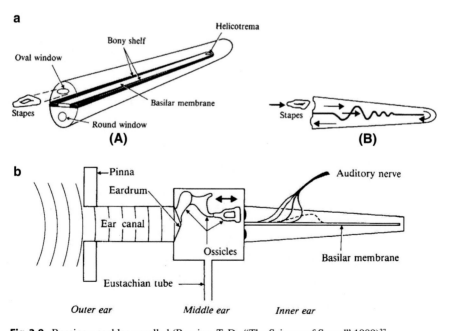

Fig. 3.8 Rossing - cochlea unrolled (Rossing, T. D., "The Science of Sound" 1989)[37]

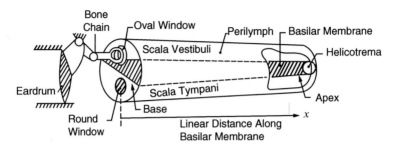

Fig. 3.9   Cochlea unrolled (Flanagan, J. L., 1972)[38]

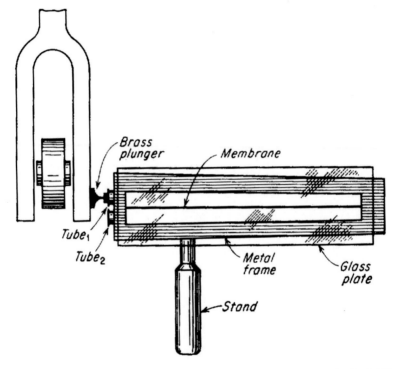

Fig. 3.10   Bekesy - cochlea unrolled (Georg von Bekesy, "Experiments in Hearing," 1960)[39]

probably the most studied element of ear physiology. In almost every case in the literature, the cochlea is straightened out ("idealized" in one text) to better understand its structure. While this provides a good visual presentation, in the long run this model may be enforcing some unsound orthodoxy about the cochlea, aggravating our confusion about how it works.

The common model of the cochlea is a tapered tube divided by a membrane along its length, with an "oval window" at the input end and a smaller "round window" as a pressure relief diaphragm at the other end. This model makes sense, the physics of fluid dynamics and wave theory work well in it and it has long provided a robust playground for mathematical analogies. (The papers describing the various subtleties of mechanics and fluid dynamics derived from this model look like spilled spaghetti to anyone not familiar with the language of calculus.[40])

The cochlea is hard to study in vivo because of its size, its limited accessibility, and its role in the organism, so a large proportion of the studies about the cochlea are theoretical and based on the conical model. This uncovers one of the shortcomings of building on these theoretical models: The results are comprehensive—they further our understanding and they seem to imply a reasonable approximation of the real thing. But the basic flaw with the model and all of the beautiful research devoted to it is that the cochlea is not a tapered or conical tube, it is a spiral. This spiral shape is not just an adaptation to provide a way of folding a tube up into a manageable envelope, the spiral itself represents an important context of proportionality. The "idealized" forms in the literature are used to simplify the model so that we can grasp it with our minds. But this simplification ignores the very shape that serves the nearly incomprehensible function of its true form—which is to precisely interact with all of the acoustical energy that stimulates it.

The drawbacks of the simplification are clear: If you take the "tapered tube" model and slice it across its axis, the cross section will be a circle. The centerline of this circle will be equidistant from all radius points out to the outside perimeter. If you slice it obliquely, the radial dimensions will be symmetrical along the mirrored sides of the ellipse, with the mirrored points along the sides equidistant. On this oblique section, the progression from the shortest to the longest radius will be a linear mathematical function. These shapes conform to mathematical and conceptual tools which are easy to work with—which are the primary reasons for the simplification of the model. All of the complex calculations of viscosity, propagation delay, tissue permeability, membrane impedance, sterocilia reluctance etc. are easy to sort out in the mathematically predictable envelope of a tapered tube.

The cochlea on the other hand is an irregular shaped in section, not cylindrical. It is also not a straight taper, rather it is formed in a spiral based on "phi," the "magic ratio" or "golden mean" found almost everywhere in nature (from shell forms and electron spins, to the sinuous arms of spinning galaxies). "Phi" is an irrational number expressed most graphically by the Fibonacci series and the familiar nautilus shell expansion.[41] One beauty of this in form is that if you slice it across its axis, the geometric centerline will not be equidistant from any other radius point around the perimeter. If you slice it across any section, the same law holds true, so what you have is a form that provides an infinite amount of possible proportions along any linear axis, across any intersecting plane.

Multiplying on this simple, but infinite elegance is the fact that sound waves will not travel down this spiral form in straight lines, rather the pressure gradients and particle motions induced in the cochlear fluids will be coerced into a vortical form. This vortex will also tend toward a "phi" spiral, but longitudinally along the axis of the cochlear spiral. This creates an intimate intersection of two

Fig. 3.11 Cochlear and
conical sections. Conical
section: Center line
equidistant from the outer
diameter (geometrically
simple). In the cochlea
section an equidistant
centerline is geometrically
complex

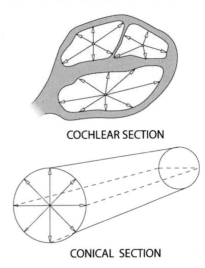

COCHLEAR SECTION

CONICAL SECTION

Fig. 3.12 Cochlear Vortex.
Compression waves in a
"phi" spiral will tend toward
vortical motion

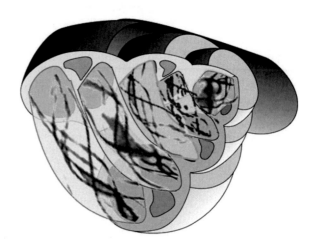

infinite or "irrational" forms; one stationary and one dynamic—conferring the infinite possibilities of our acoustical environment into the infinite possibilities of neurological activity.

If this seems difficult to get your mind around, don't fret, because it is. Given the inherent complexity, I am not going to attempt to straighten it out here, except to propose a revised physical model of the cochlea as an organ that is in the form of a phi spiral which induces vortical transfer of energy along its axis.

This spiral envelope is divided along its axis by a membrane, called the "basilar membrane" that runs the entire length of the cochlear shell dividing the spiral into

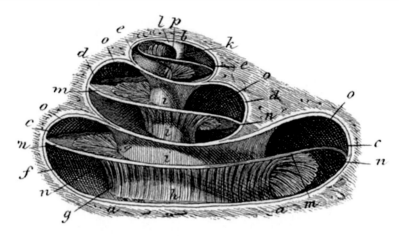

Fig. 3.13 Spiral auditory nerve system within the cochlea (Courtesy of AnatomyAtlasas.org)

upper and lower chambers or canals.[42] In the classical model the basilar membrane mechanically resonates along its length as a dimensional product of frequency; the longer wavelengths reaching farther along the structure than the shorter wavelengths.[43] This dimensional attribute accounts for our pitch discrimination; with lower frequencies exciting nerves that are deeper into the structure and higher frequencies exciting nerves closer to the beginning.

The upper surface of the basilar membrane includes the "organ of Corti"—the cilia, hair cells and nerves that sense acoustical energy. This organ lines the "floor" of the upper chamber, which is further divided by another membrane, called the "Reissner's membrane." Only two cells thick, the Reissner's membrane is stretched taught like a drum skin above the entire length of the organ of Corti, separating it from the fluid movements of the upper canal. The Reissner's membrane serves as a divider of fluids, but also of electrical potentials,[44] a feature that helps amplify its effects on the organ of Corti by way of fluctuating electrical potentials induced through its movement.[45,46]

Until recently it was thought that the Reissner's membrane moved sympathetically with the basilar membrane, but recent advances in laser interferometry reveal that this membrane moves independently.[48] As a "drumhead," the Reissner's membrane integrates the particle motion and the pressure gradient motion of the upper canal fluids translating them into a pressure gradient energy. So the Organ of Corti nerves are stimulated from the bottom by wave actions of the basilar membrane (to which they are attached) and stimulated from the top by fluctuating electrical potentials and dynamic pressure gradient energy from the Reissner's membrane. It seems that the cochlear nerves have plenty of stimuli to integrate into sound perception.

But again, this is not the entire story. It seems that the cilia of the Organ of Corti are also active, in that they respond to various conditions by producing their own sound. (Yes, the ears sing as well!).[49] There is a natural mechanism of "otoacoustic emissions" that is a part of our hearing. This internally generated sound is produced

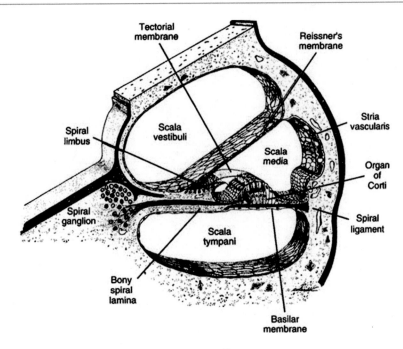

Fig. 3.14 Cochlea section (source: Davis, A., 1962)[47]

from within the cochlea, responding to and biasing the incoming sound to accentuate the signal. Otoacoustic tones are excited by acoustical energy that assists in the
capture and perception of weak or momentary signals.

The show-stopper for me is the fact all of these complicated geometries—all of
the precise acoustical transformations, the singing of cilia, the electrical activity and
wave actions of membranes, and the pressure actions of fluids—in sum, the transformation of the entire acoustical world that surrounds us into neural activity – all
occurs within a little spiral shell that is about the size of a pea.

Outside the Cochlea...

I'm going to extract ourselves from this structural examination of the cochlea at this
point. It is an "organ of infinite possibilities" and has warranted the lion's share of
hearing research—and will continue to do so. I hope that more work will examine
its vortical form and the consequent tendency of vorticity of the system. And now
that the Reissner's membrane has established independence, I'm sure that it will
also garner more attention.

Of course the transformations in the cochlea are not stand-alone transactions;
they need to be coupled with the outer acoustic world. The contact point is on little
membrane called the "oval window" which sits under the foot of the tiniest bone in

Fig. 3.15 Pea-sized
human cochlea and inner
ear structure (source:
Kenneth Garrett/National
Geographic Stock)

the human body, called the "stapes" or "stirrup" (because it looks like a saddle stir-rup). The stapes is the third in a chain of ossicles connecting the oval window (win-dow into the inner ear) to the ear drum (exposed to the outer world). The oval window is about 1/30th the area of the ear drum, so if it was directly exposed to the outer world without the benefit of the ossicles and the ear drum, it would only con-fer about 2 % of the available acoustical information into the cochlea.[50] Given the fact that airborne acoustical energy is already feeble, a scant 2 % exposure would present an even greater challenge to perception. Fortunately the bony mechanics of the middle ear provide as much as 25 dB of mechanical gain between the ear drum and the oval window.[51]

The ossicles of the middle ear—the malleus, incus, and stapes (the hammer, anvil, and stirrup for those into the blacksmith metaphor) account for this mechani-cal gain, but they are more than a simple lever system for impedance matching between the tympanum and the oval window. At the oval window site, the oval foot of the stapes sits neatly on top of the oval window, but it doesn't just move in and out like a piston, it also tilts and swings a bit (frequency and volume dependent), imparting complex phase/gain information into the cochlea.[52] Given that we local-ize sound sources by way of time domain (phase) information, this nonlinearity would enhance time domain information already present in the sound, helping us discriminate the subtle cues of sound source location. This tilting and rocking prob-ably also plays into the vortical motion of energy in the cochlea; if the action were just piston-like, it would require some throw to converge into vortical movement—a throw distance that is not readily available in the dinky little cochlea.

The action of the ossicles are also not linear with respect to volume; their motion and flexibility is mediated by muscles (like the tensor tympani mentioned above) that respond to loudness. These come into play when noises rise above about

75 dB$_{SPL}$[53]—the range where the potential for ear damage starts kicking in. 75 dB$_{SPL}$ is about as loud as you would want to hear someone speak—or where speaking ends and yelling begins. It is also about the tolerable noise level of a large crowd in a bar.

The muscle contractions provide a dynamic limiting of the ear that takes only a few milliseconds to kick in, though cautiously, it takes considerably longer to kick out. When we are subject long stints of to volume levels above 75 dB$_{SPL}$, these muscles are subject to auditory fatigue. At higher levels, such as in an amplified concert (or with the potential sound levels from personal headphones) these muscles will work really hard to protect our ears. They were not designed to handle long duration noises above the yelling level, so in doing their duty they can get really exhausted. Once the loud noises cease the muscles can go into spasm, causing dulling of sensitivity (temporary threshold shift) that will last a matter of hours to a matter days depending on how long and how loud the noise was that we were subjected to.

This threshold shift is often accompanied by ringing if the trauma was particularly loud, or aggravated by other factors such as pharmaceuticals, excessive alcohol consumption, or high-impact bouncing around. (High impact aerobics with loud music has permanently damaged many good ears.) The ringing—called "tinnitus," may be an accentuation of normal otoacoustic emissions or an unrelated mechanism, but if you have experienced it you know that it can really be a drag. Hopefully after a given time in a quieter environment, the ringing goes away. Unfortunately this is not always the case, and some poor souls are continuously subjected to a ringing in their ears as a result of high volume ear damage or a short circuit in the feedback process that otherwise shuts down the ringing once we are acclimatize to a quieter environment. Sometimes the mechanism that causes this ringing can be extreme, oscillating the tympanic membrane like a drum head to the extent that the ringing can be heard by other people.[54] (A prayer of silence for those afflicted with this problem.)

Returning to the middle ear; under normal circumstances the ossicles of the middle ear provide up to 25 dB of gain. But given the difference in surface area between the ear drum and the oval window, we still need about 5 dB$_{SPL}$ more gain to catch up with the available signal at the ear canal,[55] and it would be nice to have a little extra for those quiet nights and whispered sweet nothings.

Fortunately, this was figured out ahead of time, and mammals (other than ocean and water dwelling mammals) were provided with sound collectors which we call "ears."

The outer ear is often what we imagine when the ear is mentioned. In whole it is called the auricle, but is often divided into the outer perimeter called the "pinna," the large inner concavity called the "concha," the little bump that extends out a little just in front of the ear canal called the "tragus," and the soft lobe that hangs down below. The larger role of the auricle is to collect sound into the ear canal and gather it toward the ear drum. The gain of this whole system is not linear—that is, low frequencies and high frequencies have less gain than mid frequencies—there is as much as 20 dB additional gain in the mid-band.[56] This mid-band sensitivity conveniently coincides with the frequencies of the consonants in human speech, yielding the highest sensitivity in the band that we use for articulating words. Given the importance of human communication, increased sensitivity in this band is no surprise.

The auricle, with its graceful whorls and folds is cute enough just to hang there looking coy, but these contours are there for a reason, because the concha, pinna, tragus, and lobe all modify or tailor the soundfield as it approaches the ear canal. The first modification is amplification by way of sound collection; but if amplification was our only requirement, we would just have funnels attached to our heads focusing the sound directly on to our ear drums without all of the cute convolutions. The auricle does focus the primary sound to the ear drum, but the sound also bounces off of the pinnae convolutions in a very useful way. Sound coming from directly in front of us will bounce off of the wall of the concha and enter the ear canal just a little later than the direct sound, and because the concha is an irregular radius around the ear canal, this time delay will provide the ear with a phase-modified signal dependent on the elevation of the source.[57] (You may notice that the curves of the pinna and concha also suggest the "phi" curves mentioned earlier in this chapter.) This feature allows us to locate the height of a sound source with an accuracy of a

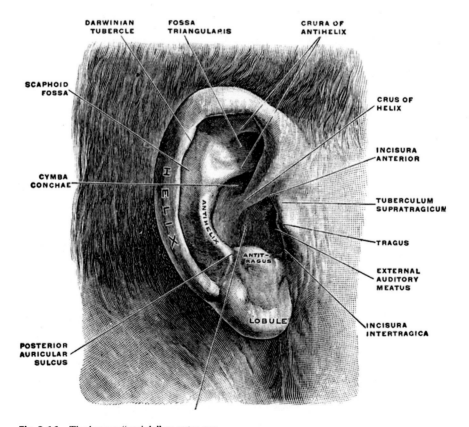

Fig. 3.16 The human "auricle" or outer ear

degree or two. This localization information is derived from the minute time delay—or "phase distortion" provided by the different path lengths between the direct and reflected sound. Due to the sensitivity of our ears to time domain information, the convolutions even allow us to locate the height of a sound source with just one ear.[58]

With this level of sensitivity, any whorls and bumps within the compass of the pinna will have an effect on what we hear, and specifically how we localize the sound source. Assisting this is the tragus, which sits directly in front of the ear canal. This looks like just a cute little bump, but it cleaves the sound coming from in front of us into two pathways. It also directs the sound bounced off of our concha into the ear canal. The tragus is small, but if we didn't have it, these sounds would fall directly into the ear canal, making our sense of sound localization akin to driving an old car with a loose front end.

All of the auricle features of the described above are formed of semirigid cartilaginous material covered with a fine-pored skin; flexible and reflective, but stiff and resilient. These characteristics are true of the whole auricle except for the ear lobe, which is soft and pliant. Again, a cute structure with the added sensual advantage—at least for some folks—of being a bit erogenous. I have had a hard time finding references to the ear lobe in the audiology literature. This is probably due to the fact that it doesn't modify the soundfield so much as it acts as a vibration damper for the rest of the auricle. It prevents the pinna from flapping in stiff breezes or vibrating in loud, low frequency soundfields. As the lobe stabilizes the auricle to allow for the clear and concise hearing, there may be some poetic connection to the Chinese association of long ear lobes with wisdom.[59]

The gain provided by the outer ear at communication frequencies explains a lot as to the priority we give the mid-frequency bands. We can resolve scads of activity in this range, so we use it for speech, music, and audio alarms. If our sensitivity band was an octave lower, our ears would need to be much larger, like elephant ears, to accentuate the lower frequencies. We would also need to speak more slowly and stand farther apart to integrate the longer wavelengths. Conversely, if our sensitivity band was much higher, our ears would be much smaller, we would probably be speaking much faster, and our languages would have more "edges" to them; we would probably use sharper and more frequent consonants—chattering like bats or dolphins.

It's not that these other bands are unimportant to us; we need the higher frequency sensitivities for sound localization, and we need the lower frequencies to gauge size and consequence; it's just that most of the acoustical data that informs our conscious decisions are bracketed by the mid-frequency band of ~250 Hz to ~2,000 Hz.

We don't require more high frequency gain because we don't really derive serial information from this band, rather we derive context information with cues about proximity, activity, and motion—these cues play more into our autonomic nervous system and our subconscious sense of placement. Higher frequencies are like the seasonings in the soup; a little goes a long way so we don't need much to savor the effect.

Low frequencies on the other hand are fairly important for serial information; we gauge size, and thus potential threat levels from large low frequency sounds. These sounds also physically lock us in to our surroundings—we hear them, and we also feel them in our bodies. Our chest cavities and guts resonate with low frequency sound; we feel it in our feet and in our bones. The fine hairs on our face and arms are also moved by the dense energy inherent in "big bottom" sounds. So while our pinnae do not collect these long wavelength sounds well, we feel the high energy density of low frequency sound through tactile and bone conducted pathways, which more than accommodates for our ears' lower sensitivities to these tones. We excite ourselves with dance music that contains lots of low frequency energy—from Native American Pow-Wow drums and African ceremonial and telegraph drums, to Bootsy Collins on his funky-finger bass (or the big bottom found in any reputable dance scene).

Dance music is a good springboard for discussing the physical and associated neural pathways for sound perception that are not specific to our ears—bone conducted hearing. The acoustical energy that passes from our bones into the cochlea is usually only mentioned relative to sound conduction through the skull, and is specific to hearing with our ears. In airborne conditions the sound level through this pathway is 24–60 dB lower than the same sound through our ears,[60] but this is not the sound that excites us to dance, rather it is the dancing music that we "hear" in our bodies that moves us so deeply. When our skin, flesh and bones are stimulated by music, we are induced to move our bodies within the soundfield. The ear is not the sole fulcrum of our movement, our center of movement is more toward the center of our bodies—the heart, belly, and hips.

The association of music and body movement reaches way back into our visceral sense of sound and consequence. Music, and its consequence on the body stimulates so many pathways in both receptor and response systems, that the literature is cluttered with ambiguous studies and unresolved questions about why there is so much neurological activity dedicated to musical perception. Part of the confusion devolves around the belief that our musical and kinesthetic senses are "non-essential" to our Darwinian struggle for survival and thus questionable in their purpose.[61] I won't argue with this rather boring conceit.

What is clear is that body conducted and kinesthetic sound perception play into the sense of self and placement in ways that are not solely dependent on auditory processing, and that multiple auditory pathways exist which excite our brains and our bodies. These areas of activity are not exclusively cognitive, but play also into the limbic (emotional) system and other areas of the sympathetic and autonomic nervous systems.

This perspective opens up the sound perception model to include ways that acoustical energy stimulates our bones and flesh, helping us integrate our bodies into our surroundings in ways that are not dependent on our ears alone. This presents a fairly loose fitting suit which we can call our "sense of hearing." Our bones and flesh, the cavities of our abdomen, our lungs and belly, the surface of our faces and our genitals—as well as our ears, are all subject to acoustical stimulation.

Our sensitivity to acoustical energy through multiple pathways is bracketed by our perceptual requirements and practical needs. It ranges from the quietest sounds we can perceive (conveniently just above the noise level of our own blood circulation), to the threshold of pain—just below the level of instantaneous physical damage to our ears. In numerical terms this range is between 0 and 120 dB$_{spl}$ (re: 20 µPa) or one million to one ($20 \cdot \log 10^6 = 120$ dB). Expressed in geographical terms, this would be equivalent to being able to simultaneously see a young child and the entire state of California, or your middle finger and the city of Los Angeles.

While our auditory pitch discrimination resides between 18 Hz[62] and 18 kHz (the top end a bit higher for a healthy baby, a bit lower for an elder), our pitch discrimination is really only accurate in the useful musical and conversation range of 100 Hz–5 kHz.[63] Nonetheless, we are affected by infrasonic energy in the range of 1–2 Hz, alerting us of seismic activity of earthquakes and avalanches (which we can feel as autonomic anxiety), and we are also able to resolve interaural time differences down to the 2 µs range (reciprocal of 500 kHz),[64] helping us discriminate transient and impulse information imbedded in the sounds that we consciously hear. (These high speed transients give us subtle time domain cues that help us localize the source of a sound.)

With this we can summarize that the extents of our acoustical energy perception lies between the time domain bracket of ~1 s and ~2 µs, and between the amplitude levels of the quiet murmur of our hearts and the thundering noises of Def Leppard on stage.

As comprehensive as this foregoing survey may seem, it really only sketches the textural range of possibilities in the vast enterprise of human sound perception.

American composer John Cage brought his puckish humor to the classical music tradition, introducing the formal idea of "silence" into the Western musical lexicon. His seminal work 4' 33" introduced the idea that a time frame—in this case 4 minutes and 33 seconds—is enough of a context to constitute "music." This piece was executed in silence by the performer, but included all of the sounds that occurred in that time; the breathing of the audience, the sounds of their initial reactions and their internal monologues, the sounds of the concert hall, the rustling of programs, the sounds of the city outside—in sum, 4' 33" presented the unfolding of time marked by the sounds at play during the piece.

Cage once related his experience of being in an anechoic chamber wherein all sounds are silenced by deep sound absorbing materials and structures and devoid of all incidental and reflected sounds. In this "silent" chamber he heard a ringing or humming in his ears. The acoustical technician explained that this was the sound of his own nervous system and blood circulation.

I was doing some acoustical product evaluation at the anechoic chamber at California State University of Sacramento. The chamber was a reasonable size;

5 m x 7 m x 9 m, and the walking surface was a taught, acoustically transparent net suspended over a sound absorbing chamber base. When I first walked into this chamber I almost fell over with a sense of vertigo, as I had not only walked into a silent realm, it was also an enclosed space without any acoustical references to space that even the scantest sound reflective surface would yield. This took some time getting used to.

Sound Menagerie: Other Animals' Sound Perception

4

> *Call the dominating inhibitions that determine our point of view anything you wish. They affect all of us, including scientists. All are saddled with heavy linguistic, national, regional, and generational impediments to perception. Like those of everyone else, the scientists' hidden assumptions affect his or her behavior, unwittingly directing thought.*
>
> (Lynn Margulis, "Symbiotic Planet")[1]
>
> *...with few exceptions, the vocalizations of animals—the songs and calls of birds, the grunts and roars and purrs of mammals, the croaks of toads, and frogs, the cricket's chirp, the astonishing array of underwater clicks and booms emitted by fish—all of these sounds evolved for purposes that had nothing to do with mankind. They bespeak other business, and it is only relatively recently that scientifically minded people have begun to pry effectively into what that business is all about.*
>
> (Eugene S. Morton and Jake Page, "Animal Talk: Science and the Voices of Nature")[2]

Given the superlative range of human sound perception, and the wondrous ways we use our sense of hearing, one might assume that human sound perception represents the pinnacle of bio-acoustic adaptation in nature. But the success of our species is not in specialization; humans are generalists, and while we hear quite well, our hearing is by no way "the most," the "best," or the "finest" of all creatures—it is just very well adapted to our habitat and our perceptual priorities. The impressive quality of our gift has unfortunately set much of our scientific research up to compare other animals' use of sound to our own. While this strategy has yielded some amazing information on animal bio-acoustics, we have by the limits of our own perception been hampered in understanding some of the finer points of how other animals use sound. This is particularly the case when those ways are alien to us.

M. Stocker, *Hear Where We Are: Sound, Ecology, and Sense of Place,*
DOI 10.1007/978-1-4614-7285-8_4, © Michael Stocker 2013

Many of our studies of animal perception have been framed under broad assumption that animals mostly use sound for communication. This has established classifications of "sound specialists" and "sound generalists" depending on (among other characteristics) an animals' ability to vocalize. This has given short shrift to animals that are finely adapted to hearing but who don't happen to talk about it.

Unfortunately even when the animals can speak, we are unable to ask them simple questions and get straight-forward answers. As a result, studies of animal sound perceptions are necessarily full of speculative language and broad assumptions. This situation is aggravated by the anthropocentric perspectives we use when we set out to discover something about other animals' behavior and characteristics. This isn't really short sighted in light of the fact that most of our exploration into the animal world is driven by our desire to know ourselves, so framing our discoveries in human terms is accepted practice. But when we use this information, we must bear in mind the foundation of our assumptions along with meaning of the data. Because while it is likely that much of the groundwork has been laid for understanding the mechanisms of animal sound perception, we are still only scratching the surface as to the range of their sensitivities and even how any particular animal uses sound.

Given our anthropocentric framing and assumptions, our understandings of animal sound perception have been founded on some fairly blunt tools. By-and-large, bio-acoustic research is limited to only a few accepted scientific methods: evaluating trained behavioral studies in captive settings (the Skinner/Pavlov method); examining laboratory induced neural-electrical responses to acoustical stimulus with Auditory Brainstem Response (ABR)[3] and later Auditory Evoked Potential (AEP); or observing animal responses to acoustical stimulus in their own habitat.

There are obvious drawbacks to each of these methods. In laboratory settings captive animals are subject to stresses or biological constraints not found in natural settings. Lab animals are also hard to train, and don't necessarily behave in natural ways when they are trained to associate acoustical stimulus with punishment or reward. Auditory Evoked Potential (AEP) attempts to bypass the training problems by measuring neuro-electrical activity evoked by acoustical stimulus, but animals harnessed into an AEP jig are often sedated and may be traumatized in ways that could unpredictably influence the tests.[4] Wild animals in natural settings, on the other hand, are in complex relationships with their observers and their environment, which can yield ambiguous results to any programmatic tests.

I have also found that the technical orthodoxies of testing procedures can limit the usefulness of the test results. For example, the human priority of pitch discrimination leads researchers to perform frequency sensitivity tests—or "audiograms" on animals. Typical audiograms use sine waves to determine specific frequency sensitivities, but sine waves are "pure tones" not commonly produced by animal vocalizations[5] and may behave in neurologically unpredictable ways for animals who don't have a need for accurate pitch discrimination.[6] Some animals are distinctly *not* sensitive to sine wave stimulus[7] and reveal as much as a 20 dB increase in sensitivity if "band limited noise" is used in place of sine waves.[8] Given this limitation, audiograms taken with sine waves (most of them) need to be discounted to some degree.

This is particularly the case at lower frequencies where pitch discrimination is poor even in humans, and the perception of pitch melts into a perception of rhythm. (Humans can hear acoustical energy at frequencies below 20 Hz, we just don't hear their pitch, rather we begin to hear these "sub audio" sounds in the "time domain" as a fluttering or pulse-train sound.)

The physics of low frequency sound brings up another challenge to audiometry, and is the most common "nemesis issue" in all branches of acoustics: Due to the long wavelength of low frequencies, they are difficult to handle. Long wavelength sounds are hard to reproduce in environments that are smaller than their wavelength (the wavelength "λ" of 20 Hz is 56 ft in air). The characteristically higher energy densities of low frequencies also tend to interact with the surroundings, vibrating the furniture and fixtures and interacting with room geometries in ways that are hard to predict—particularly as the wavelengths approach the dimensions of the room. I wouldn't necessarily dismiss all audiogram results below 140 Hz (λ = 8 ft, the height of a typical lab ceiling), but I wouldn't settle for the results as gospel unless I knew something about test apparatus and the test environment.[9]

These systematic shortcomings notwithstanding, perhaps the greatest liability to any study of wild animals resides in our assumptions about who these animals are, what they do with the perceptual systems under observation—and how they behave when being observed. It may be difficult to admit, but some animal subjects occasionally prove to be cleverer than we think. A subject's long term memory, suspicions of the observers, associative reasoning and rhetorical behaviors can seriously flummox the best behavioral studies.[10] William Tavolga, in his benchmark compendium, "Hearing and Sound Communication in Fishes" relates how a subject fish was trained to respond to acoustic stimulus by jumping from one part of a tank to another. Researchers were pleased to work with this little fellow as he learned fast and demonstrated a remarkable sensitivity to the stimulus. Far into the study it was found that the fish was watching the researchers and responding not to sound, but to the motion of the scientists as they reached for the test button.[11]

Even with these occasionally revealed frailties, there is an almost evangelical trust in scientific orthodoxy. Trusting the literature is a prerequisite to scientific progress, but the pedagogy can sometimes drive the model—to the detriment of true learning. In this context, when a behavior doesn't fit a known, comprehensive model, it runs the risk of being summarized as an "anomaly" or "anecdotal," when in fact it may bear critical behavioral clues.

This became clear to me early on in my work when I wanted to know something about the hearing capabilities of mockingbirds. My inquiry was prompted by an unusual mockingbird behavior I witnessed as a youngster. A father of a childhood friend (who was really into gadgets) purchased an early version of an automated garage door opener. This device worked on ultrasound (unlike the later radio frequency models). I presume that the frequency was just above human pitch discrimination because we were unable to hear it. Nonetheless, a local mockingbird did manage to ape the thing and delighted himself with opening and closing the garage door throughout the day.

I wanted to know more about this little stunt so I called an avian auditory special-ist. Using behavioral audiometry, this scientist had been extracting audiograms from birds for 30 years. I don't want to diminish the monumental extent of his work, but he assured me that few birds can hear much of anything above 8 kHz. This didn't square with my mockingbird experience, nor does it square with the fact that many birds—including common starlings and jays—can produce dense sounds in their vocalizations in the 10–15 kHz range—well above their "known" hearing range. Nature can be extravagant, but she is not frivolous; these vocalizations are not high-pitched without reason.

I took this quandary to another specialist, who assured me that based on the long-standing reputation of the first specialist, that if he "…had determined that few birds hear much above 8 kHz, then that was the extent of bird hearing." My experience was "anecdotal" only. As to biological purpose for the broad sound spectrum of bird vocalizations, this was a puzzle that "we may eventually understand once we expand our methods to encompass the data…"

The scientific foundation of our understanding of animal hearing is largely based on "behavioral audiometry." Typically animals are trained to associate an acoustical signal with either a pain or reward.[12] Testing will start with a known center fre-quency to produce a predictable positive or negative response. The frequencies are then adjusted in pitch or volume across a range to induce the same type of response. Once the responses stop, it is assumed that the "thresholds" or range of sensitivity has been reached—with the assumption that all other sounds are outside of the sen-sitivity range. This approach is necessarily methodical, but it often systematically overlooks the possibility of secondary thresholds (sensitivities in ranges well above or below the test frequencies and levels), or behavioral responses to acoustic energy that are "collective cues" particular to the subject's behavior within a school, flock, or herd, but not specific to the individual animal. This method of audiometry also ignores the possibility that animals are sensitive to sound qualities other than ampli-tude and frequency.

Surprising secondary thresholds were discovered in shad, a schooling fish which like many fish had been known to hear only up to about 5 kHz. This high-frequency limit was established through audio testing centered on frequencies associated with their conscious activity, such as sounds of kin. It was only by searching for ways to keep shad away from hydro-electric dam intakes that their secondary thresholds were happened upon. In their explorations into sounds that would repel shad one of the researchers suggested that they might avoid the sounds of their predators.[13] Not too surprisingly it was found that shad were sensitive to the hunting clicks of dol-phins in the 40–180 kHz range.[14]

These "secondary thresholds" are in ranges that stimulate voluntary nervous sys-tem responses—triggering a conscious evasion of the sound source. But because this adaptation likely evolved to avoid predators, it also stimulates their autonomic and sympathetic nervous system responses (stress and metabolic anxiety). Nervous system responses at this level don't necessarily induce an animal to "cognitively identify" a specific threat. As a consequence sounds merely in the frequency range of dolphin hunting clicks can scare fish away from dam intakes—even when the sounds are not simulated dolphin clicks.

The fishes are also responding collectively to the threat, not as individuals. This natural condition could potentially confuse audiometric testing of a single fish that ordinarily faces threats within the safety of a school.

This line of reasoning highlights a strong scientific predilection to view animals as individual organisms or as pairs of interacting creatures that use sound competitively to establish territory, secure breeding rights, attract mates, or hunt for food—the Darwinian thing. There is ample evidence of Darwinian survival behavior in animal vocalization and sound perception, but exclusively using the "survival of the fittest" filter to frame the behavior of an individual animal is beginning to wear thin as we learn more about the role of cooperative adaptations in evolution.[15] The "competitive individual" model serves little to explain the acoustic communities represented in the synchronized stridulation of crickets, the coordinated motions of flocking birds or schooling fish, or the pulsing song of the frog pond.

The above discussions only touch on some of the challenges in animal bioacoustic research involving creatures that we can somewhat relate to—community animals with identifiable breeding behaviors. We have even farther to reckon with animals that we don't identify with and are hard to anthropomorphize, but nonetheless have adapted to acoustic stimulus—barnacles, jellyfish, lobsters and clams, for example. We can sometimes see their sound-sense in action; we can even make some informed conjectures about how they perceive and use sound, but at some point our scientific imagination falls short, as there is no way we could really grasp barnacle priorities. (Paraphrasing philosopher Ludwig Wittgenstein: "If a barnacle could talk, we wouldn't understand what it was saying.")

Given this broad caveat, there is much to learn about animal sound perception. Hopefully the following exploration will give us a toe-hold on the topic of animal bio-acoustics, and through this a little broader understanding of our own acoustic world.

Fish Ears and Ocean Hearing

It is generally accepted that all life on our planet traces its origins back to the sea. The "ocean water as amniotic fluid" metaphor is so compelling that it set the stage for early theories about evolution (e.g., the "gills" on early stage human embryos imply our primeval piscine pedigree).[16] Because the earth bears water, it bears life, and because the ocean is such a fine medium for the generation of life, it is also a fine nursery for the evolution of the senses. Thus the sea is a good place to start our exploration into the development of hearing.

At the dawn of life on this planet—when the first prokaryotes and cyanobacteria figured out how to "cook with chemistry,"[17] they didn't need much in the way of dynamic perception; they just grew, respired, and let the waters do the distribution work. But once photosynthesis and nucleated cells cranked up, there was a decided advantage to mobility—and the ability to orient. Initially orientation requirements were simple—move toward the goods: mostly light and nutrients. As organisms became more complex, other orientation requirements emerged: heading away from danger, shuffling around to the good feeding spots, and heading home for the night.

It would be hard to imagine the various orientation mechanisms that were tried then winnowed out over time. Unfortunately the first soft-tissue organisms haven't blessed us with much of a fossil record, but the living great-grandchildren of some of the early life-forms do give us clues to ancient orientation mechanisms and the successful evolution of the ear.

In the medium of salt water, most creatures hover around neutral buoyancy, unburdened by the effects of gravity and with freedom of movement in the x, y, and z axis. As fun as this sounds, orientation cues can be a challenge—particularly at night or in deep waters where sunlight from above doesn't penetrate (only a few hundred feet down). Jellyfish have handled this problem with a ring of "statocysts"—sense organs around their mantle, each containing a "statolith" or little stone mass responding to gravity and cueing the jellyfish into which way is up.[18] Statocysts are found in many invertebrates—even lobsters and crayfish, and are probably one of the first cuts at what has evolved into the semi-circular canals of vertebrates, with "otoliths" suspended in a fluid-filled sensing envelope.

Fossil records of the inner ear start appearing from out of the Devonian period—300 million years ago—in the remains of strangely shaped creatures; spined and armored, but jawless and toothless that swam the vast, warm, primordial oceans. One family of these creatures called the "Ostracoderms" is the oldest known group of animals with backbones. Among the ostracoderms were some who left behind fossil impressions of whorls of bone in their skulls.[19] These whorls resemble the semicircular canals that comprise the balancing organs in contemporary ostracoderms: the hagfish and lampreys. In lampreys, these organs also show sensitivity to vibration and low frequency sound[20] lending us the first suggestion that hearing and orientation serve an intersecting biological purpose.[21]

Life liberated from the effects of gravity would naturally encourage the development of sophisticated organs of orientation, but water is also a wonderful medium for sound transmission. Being 3,500 times the density of air, and noncompressible, water couples and transmits sound very efficiently. Considering that light transmission is significantly reduced in water—in terms of both turbidity and light absorptivity (making visual perception unreliable), ocean conditions are ripe for the evolution of a magnificent array of hearing mechanisms.

If we start with the basic body of any animal, we find that soft tissue (being primarily made up of water) has an acoustical compliance close to the acoustical compliance of water. If this tissue includes some sort of physical sensing system, this "sense of touch" could serve as a rudimentary sense of hearing. Sea creatures have taken this good thing and refined it in many ways.[22] The lateral line of many fishes is one of the more familiar examples. This organ runs along the side of most fish and is comprised of the same type of hair cell and nerve structure or "cilia" found in the inner ears of humans and other terrestrial animals. In fish it is stimulated by particle motion, helping the fish localize specific sound sources,[23] but it is also stimulated by

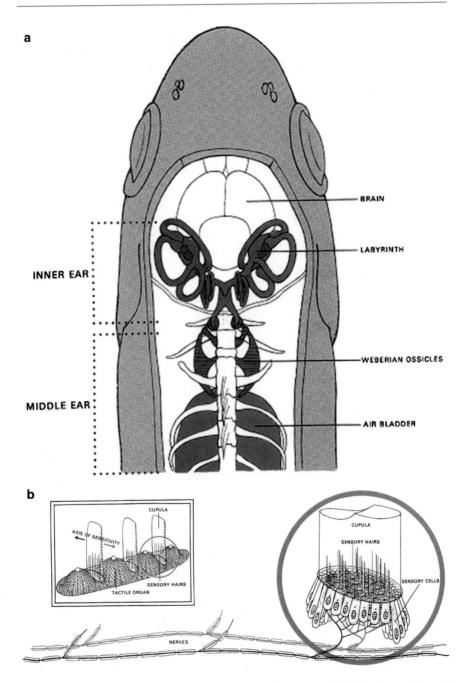

Fig. 4.1 (**a**) Internal hearing system of teleost fish showing the labyrinth, Weberian ossicles, and swim bladder. (**b**) Lateral line (Time-Life pub.)

pressure gradients and is sensitive to the pressure waves generated by adjacent fish in a schooling formation.[24] This pressure-gradient sense accounts for how a school of fish moves in concert, avoiding approaching predators and deftly navigating through the water in ethereal cloud-like formations almost as a single body.

Many fish also have an internal organ called a "swim bladder." This is usually situated centrally, internal to the animal, almost as if the body is wrapped around it. The swim bladder serves many purposes depending on the species. It is filled with gas, so it mediates buoyancy, but it also acts as a resonant bladder, resonating like a balloon to external sources of acoustical energy. In many fish this "balloon" is coupled to the fish's inner ear by way of a set of bones called "weberian ossicles"— much in the manner that our own ear drum is coupled to our inner ear through the little bones in our middle ear. The action of the ossicles, in conjunction with other acousti-sensory inputs gives fish an acoustical grasp on their surroundings.

While fish respond to a wide range of acoustical stimulus, most fish audiograms show poor pitch discrimination. Pitch discrimination is a sound sense attributed to our mammalian cochlea—which fishes don't have. But pitch is not necessarily a critical sound characteristic for fish. In their habitat, sensing proximity, sound source direction, rate of change, or phase relationships between particle and pressure gradient information are far more important sound characteristics.[27] Thus it is likely that in lieu of pitch cues, the fishes' acoustical sensors respond to textural, time domain, and acoustical density (characteristics of sound that are low priorities for humans).

For example, some fish need to discriminate the minute perturbations in their local soundfield while swimming in chaotic and loud water currents. These conditions are met when a trout swimming in a frisky brook locates and captures a caddis fly that has touched the top of the water.[28] This scenario would be akin to our being able to speed around on a freeway in our convertible Mustang with the top down, and hear a butterfly land on the bumper. From an acoustical standpoint, the fish's turbulent environment should overload and mask the sensitivity of their hearing organs. But the trout has a complex way of deciphering these extreme acoustical dynamics. In this setting pitch discrimination is not very important.

There are many properties of water that engender sensual realms outside of our perceptual grasp. Water is not as homogeneous as air; it has density and pressure gradients that vary widely with turbidity, turbulence, salinity, temperature, and depth. You might imagine an underwater environment as a rich mélange of blending densities produced by the motions of eddies, currents, and tides. These swirling nuances of density affect the transmission of acoustical energy in water, giving aquatic animals cues to the current flows, temperature, and chemical characteristics of their surroundings expressed in its dynamic acoustic qualities.

In this inhomogeneous world, swimming animals leave current trails behind them that tell of their recent passage through the medium. It is likely that some fish can sense these currents through pressure gradient cues much in the manner that we can see bio-luminescent fish trails in the ocean on dark summer nights. It is known that harbor seals can sense these pressure gradients and density disturbances—by way of their extra long whiskers. The seal's whiskers are fine-tuned to pick up water disturbances including the wake of fish, enabling them to home in on dinner, even in dark or turbid waters—at distances of up to 180 meters![29]

Another dimension of marine acoustic sense outside of our range of perception is the sense of "passive sonar." This is the ability for animals to determine spatial and geometric relationships of their surroundings by just listening to how the physical features of the environment affect their sound field. This is explored in some recent studies on how background ocean noise "illuminates" marine habitats much in the way sunlight illuminates our visual surroundings. This allows ocean animals to "see" their environment though sound.

One likely source of this "illumination" is the ubiquitous and pervasive sound of snapping shrimp (*Cragnon Alpheus* and *C. Synalpheus*) found in most coastal waters. If you are a skin-diver in temperate waters you know their noise as an omnipresent hissing and popping sound that sounds like sand swishing around, or a vast seascape of underwater pop corn.[30] Dr. John Potter of the National University of Singapore Acoustic Research Lab has been developing a system of sensing underwater objects using this background noise.[31] His results using passive sonar are quite clear in a phenomenon he calls "Acoustic Daylight." He also suggests that nature probably has a leg up on his work. If this is the case, it is likely that marine animals would be using the shrimp noise to "see" their surroundings and would account for how coastal ocean animals navigate, hunt, feed, and breed at night or in deeper waters not illuminated by sun or moonlight.

The idea that ocean animals use available sound to navigate and orient is expanding with studies revealing how reef animals identify and locate their "home reef" by imprinting on its characteristic sounds. Many reef animals' early life includes a dispersal and larval stage. During this vulnerable stage these small organisms head out to sea to escape the forest of hungry mouths living in the reef. Once they are large enough to survive the reef habitat they head back home using the imprinted sounds of their native reef as a homing beacon.[32] These sounds include the snapping shrimp, but also the chorusing of fishes, the creaking of lobsters, the whistling of sea urchins,[33] and the burbling of barnacles[34]—as well as the characteristic sounds of local coastal currents, waves, sand and seaweed. Like any living forest, a reef will have a colorful auditory scene unique to its setting and resident biota—a delicious and familiar soundscape that entices the pelagic-form reef dwellers back to their birthplace once they are ready to take up residence.

So far the acoustic homing studies have mostly concerned fish fry, but corals, crustaceans, and barnacles also go through a pelagic dispersal and larval stage, so it would stand to reason that these other reef animals might use sound cues in a similar manner.[35] Unfortunately invertebrate research is largely focused on adult animals because they are more stationary and thus easier to study. Larval stage invertebrates are fascinating because of the dynamic transformations that occur as they grow, but their evanescent forms make them poor behavioral subjects. Their "shape-shifting" also complicates any study of their perceptions because their transformations probably include transformations of their perceptual mechanisms as well. So with most of these lowly creatures we just don't have much data.

We do know that adult corals, anemones, barnacles and other stationary hunters have proprioceptors to help them intercept their mobile prey. We also know that most of these sea animals have some form of mechanoreceptors in their tissues and muscles that mediate response to dynamic environmental conditions. But our understanding of how their organs respond to acoustically induced pressure gradient or particle motion energy is at best rudimentary.[36]

While we do know more about the sound perceptions of marine vertebrates, we are still far from understanding the breadth of the field: of the estimated 25,000 species of marine vertebrates, we have hearing data on less than 100 animals.[37] Most of our research has literally been splashing around in shallow water framed by our own perceptual priorities. We live on a "blue planet" defined by her oceans, which are vast beyond human comprehension in many dimensions. Flying above the ocean we perceive it as an expansive field of water, stretching off over and beyond the horizon line. Sailing across the ocean's surface, its breadth is immense and humbling. With the added dimension of depth, its scope is even more incomprehensible. The ocean covers 4/5th's of the planet surface and at an average depth of 13,000 ft, or 2½ miles. Unlike terrestrial life (which is bound by the thin layer of soil beneath our feet and up into the air within the compass of our eyesight), the entire water column—from below the sea floor to splashing waves is host to life; suspended where opportunity allows and nature directs. Obscured from our sense of vision, but saturated with acoustical energy.

Whale Details: The Hearing of Whales and Dolphins

Sailors have long spoken of the songs of the sea—the Sirens, Nymphs and Nereids who seduced or serenaded them across the waves. Often these accounts were dismissed as madness induced by the sailor's endurance of long and lonely stretches over the silent seas. It has only been in the last half century, as the military began to probe the deep waters with their own sound that the mysterious sea sounds have emerged into scientific inquiry and revealed to be the songs of the whales.[38] But recognition of the musicality of cetaceans dates back at least as far as the seventh Century B.C.E., when dolphins of the Aegean Sea recognized Greek musician Arion as kindred soul and rescued him when he was cast into the sea by his thieving shipmates.[39] This tale speaks to our affinity towards the singing cetaceans which is perhaps why we have more information on whale and dolphin hearing than any other ocean species.

Cetaceans come in two flavors: the Odontocetes, or toothed whales including the dolphins, porpoises, Orcas, beaked, and Sperm whales, and the Mysticetes, or baleen whales, including the Humpbacks, Gray whales, the Blue whale, and the other rorquals. ("Odonto" refers to "teeth;" "Mysti" implies "moustache," referring to the baleen.)

The early whales—the amblocetus, or "walking whale"—were once terrestrial animals, but they tossed in the towel on the land-lubber life, returning to the sea some 50 million years ago during the Eocene period. It was also at this point that the apparent rift in the family occurred. The anatomical similarities between the families—elongated bodies with flippers and fins—are due to their common habitat, but the differences between these two sub-orders of cetaceans are as distinct as the differences between horses and cats. As is the case with the cats and horses, the Odontocetes are hunters; the Mysticetes are foragers and grazers.

Acoustically the distinctions are just as clear, with adaptations that reflect their respective feeding strategies: The Odontocetes mostly use high frequency sounds—an adaptation that reflects their hunting and close range communication needs. They locate, range and fix their fast moving and evasive prey with high frequency sounds. Except for the sperm whales, Odontocetes hunt cooperatively in packs, so they probably also use high frequency vocalizations to communicate and synchronize hunting strategies in close proximity to their kin. High frequency vocalizations lend to these purposes well, with short wavelengths that yield fine sonar details and high density data streams. Mysticetes on the other hand mostly use lower frequency sounds for long distance communication and social behavior.[40] Mysticetes forage for plankton and krill, so it is probably more important for them to compare notes on food abundance at long distances with their kin. Long wavelength sounds lend to this, traveling over hundreds to thousands of miles of open ocean. There is also evidence that Mysticetes use their low frequency vocalizations for long range echolocation—bouncing their sounds off of deep ocean geological features such as seamounts and ocean ridges in their long migrations across the sea.[41]

Most of what we know about all cetacean hearing comes from the captive studies of dolphins. This is largely a matter of convenience because dolphins are a manageable size. A few species are even quite gregarious and amenable to human companionship and training, which makes them even more appealing as a study subject. A 50 ft, 40 t whale is quite another story; difficult to handle and not particular to human companionship. As a consequence there are some large gaps in what we know about Mysticete hearing, so audiograms of large whale are surmised from what we know about their smaller kin, merged with observations of their bioacoustic behavior in their ocean habitat.[42] Additional assumptions are drawn from comparative physiology, when dead whales are dissected and the mechanics of their ears are compared to the organs of other mammals.[43]

Cetacean hearing has many similarities to terrestrial mammal hearing, but there are significant differences as well; whales and dolphins don't have outer ears for hydrodynamic reasons, and they don't have tympanic membranes, probably for hydrostatic reasons,[44] so we know that they don't collect sound in the same way terrestrial animals do. What they do have is a way of conducting water-borne acoustical energy into their organs of hearing by way of a fatty lipid system in and around their heads.[45] This system is really apparent in the odontocetes, from the smallest harbor porpoise to the great Sperm Whale who all sport some form of "melon" in their forehead—a bump or protrusion filled with fatty lipids.

The melon has long been associated with echolocation and hearing, but we are only just beginning to understand some of the complexities of this organ. For example, it is not just a static "sound lens;" in the beluga whale (and perhaps others), it can be grossly deformed during phonation.[46] (One researcher quipped in a private communication that the beluga's ability to transform this melon is "mind blowing" to witness.) It is also possible that the internal densities of the melon can be modified by way of thermal regulation (through blood circulation[47]) which in combination with other cavities and bone structures helps create sound focusing channels in their heads. Even in dead specimens there is enough graduation of density in the melon and surrounding structures to enable them to produce highly focused beams of outgoing sonar signals[48] (that is, if they were alive). This same lipid material is found in channels of the lower jaws of dolphins—coming in contact with the middle ear at the back of their jaw. These serve as primary conduction paths for sound into their hearing organs.[49] (Alligators also have relatively thin, lipid-filled lower jaws, suggesting a convergent adaptation to underwater sound perception.)[50]

Body and bone conducted sound has not been studied in detail, but recently the mobility of the bulla—the bone envelope containing the inner ear organs, was introduced as a potential sound transmission mechanism as well.[51] A petrous bone envelope is common to the hearing organs of all mammals, but in most other mammals the housing is external to the cranium in a globe called a "bulla." In terrestrial animals the bulla is fused to the skull, providing a stable platform relative to the disposition of the skull, the eyes and the outer ears. In odontocetes[52] and hippos[53] the bulla is mobile, which is likely an adaptation to underwater hearing.

The vocalization frequencies of cetaceans vary depending on the species. For larger animals such as the Blue whale, they are predictably lower; from 3 Hz to 40 kHz. Smaller animals like the "common" dolphin vocalize from 1 kHz to above 150 kHz.[54] As in humans, these animals can hear well above their vocalization frequencies. For example the harbor porpoise' smallest vocalization fundamental wavelength is ~2 cm but their sonar can detect 0.3 mm diameter wire targets.[55]

The wire-thin resolution of their sonar seems remarkable enough, but these higher ranges of sonar also yield more than just surface details. At these higher frequencies, sound penetrates into soft tissue, so these critters can resolve the internal structures of their sonar targets—much in the way that medical sonograms resolve soft tissue structures in humans.[56] This allows them to hone in on the soft tissues of their prey so they can avoid starting off their fish dinner with a mouthful of boney armor.

The past 35 years of scientific research into odontocetes' hearing has yielded incredible amounts of data on these creature's auditory capabilities. But even as our scientific tools become both broader and more refined, the research papers are still as rich in speculation and conjecture as they are in information. This is even more the case with the baleen whales. Scientific data runs very thin here, so conjecture and speculation largely rule the research roost. Unfortunately for our curiosity, in addition to the size problem, none of the Mysticetes have demonstrated much more than a passing curiosity in becoming intimate with humans—and they are hard to bribe into a relationship by doling out tons of plankton and krill for them. With these circumstances, aside from examining the physiology of dead animals, we are largely left with observing them in their habitat—and perhaps abusing them with noises to see how they react. Maybe our best information will come from just listening to them and hoping that someday we may understand what their songs mean.

Quite a few years ago a young female gray whale washed up on a deserted beach near where I live. It was clear that she had run afoul of a tanker or cargo ship, having a regular set of deep contusions along her body inflicted by a large propeller. Her carcass remained unmolested by humans over the months that I visited to watch her decompose. Near the end of this vigil her hide was spread over the exposed bones of her ribs and spine, settling into the sand on an oily mat of rendered blubber. Her viscera and soft parts had long been consumed by carrion birds and maggots. Her head—proud above the pool of a carcass, was beginning to reveal its mysterious internal shapes. It was at this point that I felt compelled to reach in and get intimate with the structures of her head and skull.

Armed with a long blade knife and elbow length gloves I cut into the head. Removing the outer layer of skin and putrefied blubber, scraping back the clumps of rotting flesh I began to uncover a most miraculous structure. The cranium itself was no surprise; it was just large and resting on the two lower jawbones. At the top and at the front of the cranium were a symmetrical set of rotating bones that had served to close off her blow hole as she dove. Extending forward from the cranium above the lower jaw was a set of four long graceful bones—like the fingers of a harpist— that comprised the upper jaw and the envelope of the rostrum. These were well articulated and it was clear that movement was important to their function, as their hinges into the cranium were deeply grooved and complex—allowing for precise and delicate motion. As I carefully loosened and pulled away these bones they revealed the inner chamber of the rostrum, and like a pearl sitting in an oyster, there was a long organ within it extending from a pocket in the front of the cranium out to the tip of the nose. This organ was nesting in a long and delicate bone channel. To the best of my determination this organ was mostly fatty lipids, and about the size of a young girl's leg. I gently lifted this organ from its nest and held it up. This was a thing to behold, and even after months of being out of water, it was very much intact and fresh. In place it had connected directly into a porous frontal process of the cranium directly in front of the brain. Because of its proximity to the brain, I can safely assume from that this organ is intimately involved with neural processing. I can assume from the size (almost equal the volume of the brain), the surrounding mechanics, and the location, that it is an important sense organ. Given the role of similar materials in odontocetes, I can surmise that it probably has something to do with sensing vibration.

When I returned home I called Dr. Darlene Ketten at Woods Hole Oceanographic Institute to see if she could shed some light on this for me. Dr. Ketten is one of the world's foremost specialists in cetacean hearing and physiology; if anyone could tell me something about this organ, it would be her. As informed as she is, she didn't exactly draw a blank—she was familiar with the structure, but she was much less willing than I was to speculate about its purpose.

And so it goes with the acoustical perception of our last frontier—completely surrounding us, but still deep in mystery.

Real Ears

In the foregoing discussion, sound perception in aquatic animals was examined by way of their organs of hearing. While technically some of these organs could be called "ears," generally when we think of ears, most of us picture the pointy structures sticking up off the sides of animal heads. The word itself lends to this pointy image; a tight vowel sound pinching into narrow consonants. Human ears notwithstanding, when I hear the word "ears," the panoply of ears that parade into my imaginal view are the ears of rabbits and mules, or the floppy ears of dogs. These ears are all remarkable structures, and are all the more remarkable because they are all very different.

The differences in ear shapes reflect each animal's role in nature and are formed and adapted to each animal's bioacoustic niche. Evasive prey animals such as deer, antelope and rabbits have large ears that are almost like periscopes, rising up above their heads to independently swivel—the better to capture the delicate rustling that might betray the motion of a stalking predator. Their highly movable ears are elegantly adapted to scan the surroundings (horses have 17 muscles for each ear that help rotate and orient the pinna.[57])

On the other hand, predator animals like lions and wolves tend to have shorter, forward-focused ears that keep a bead on the motions of their prey. Domesticated dogs are all over the map and have all of the exceptions that prove the rules, having been bred for thousands of years to cultivate particular attributes (from excavating badgers out of their burrows to playing pinochle with home-bound elders). In this regard, viewing dog ears with a mind to functionality is an amusing exercise in service pedigree. "Nose hounds" may have floppy hang-dog ears as a consequence of an earlier burrowing role as fox hounds or badger dogs—the better to keep dirt out of their ear canals. Guard dogs can have perky ears that orient toward intruders, and racing dogs or fighters may have tight ears that can trim to the wind or lay back out of harm's way. Some dogs even sport ears that are clipped, bobbed and splinted to make them look "cool." Given the huge variety in domesticated canines, sorting out each of their relationships to sound would need to be taken on a case-by-case basis.

This structural variety is probably represented in each dog's listening abilities as well. What we know about canine hearing comes from studying beagles and poodles bred for lab use. Behavioral audiograms of lab dogs show their frequency sensitivities between 40 Hz and 46 kHz. Surprisingly, for an animal with a hunting pedigree their thresholds are no better than human sensitivities within the human hearing range.[58] Perhaps this reflects a domestic adaptation to the human voice.

Domesticated cats, on the other hand, have maintained the dignity of their species' God-given ears—the keen ears of predators. This suits their nature, for as lackadaisical as little Fluff seems around the house, she is still a killer—silently stalking birds, rodents, lizards and insects any chance she gets.[59] Hearing capabilities are probably more uniform among cats than dogs, suggested by their more uniform pinna shapes—and their breeding pedigree for appearance and personality rather than for dedicated service and household chores.

Fig. 4.2 Cat ears showing anthelicine sulcus and pinnae convolutions

Audiograms show that "purpose bred" cats (amenable to lab handling) are cognitively sensitive to frequencies from 60 Hz to 90 kHz,[60] at thresholds 10–25 dB lower (more sensitive) than humans in our most sensitive band. In terms of ear mobility, the cat is Queen of the beasts, with 27 muscles in each ear to change its shape, focus, and lay. This mobility is probably due to the fact that while domestic cats are predators by nature, their small size also makes them a tasty morsel-sized prey, so their ears need to serve the dual duties of prey location and predator alert. This dual functionality may also account, at least in part, for the soft little dual flap part way down the outside pinna edge that cats have.

If you look closely at a cat ear, this little structure—softer than the rest of the pinna, has a lateral groove across it that can channel sound from behind the cat directly down into their ear canal. Even when the cat's ears are laid back against their head ready for fighting, this channel still provides a clear path into the ear canal, allowing a pathway for higher frequency sounds—the sounds of fast action—to enter the ear.

For those who like adding obscure words to your vocabulary, this pinna detail is called the "anthelicine sulcus.[61]" To see it in action, reach your hand about 12–18 in. behind a cat's head and gently rub your thumb and finger together to produce a little friction sound. The cat's ears will twitch and she will almost unavoidably turn toward the source. You can do this a few times to see how well she locates behind her to the left or right (until she leaves in disgust), as she is hard-wired to respond to unseen sounds behind her.

This sulcus may also serve as a damping structure, somewhat like the lobes of our own ears. The cat pinna is a thin, taught cartilage that would otherwise resonate like a drum head, so this little dual-fold soft patch located where it is would be a good vibration damper for the whole pinna.[62]

Unlike other predators, human pinnas (like those of most "higher primates") are rather flat things lying along the sides of our heads. They do not really approximate the collecting horns of other mammals. You could say that the primate ear is a "generalist's ear" and not finely tuned to gather specific frequency information; rather our ears lend to an accumulation of "more general" sound cues. This is not the case with most other terrestrial mammals, which have more cup, shell, or collecting-horn shaped pinnae, with more pronounced convolutions and textures within their deep outer ear. The cat pinna is a case in point, with deep convolutions and whorls surrounding the ear canal.

As in humans, these whorls set up time-domain interference within the ear—standing waves and reflections that distort the auditory soundfield in useful ways, accentuating the temporal-spatial distinctions of the incoming sound. The cat's deep textures would profoundly affect the time domain information of the incoming signals, giving cats an advantage in localizing sound sources. The cat's deep pinna convolutions are probably adapted to the important environmental acoustical cues that cats need, rather than the social, vocal-linguistic cues that humans depend on. Dogs also have these deep pinna convolutions, and I suspect that lions and tigers do as well, though I haven't personally checked.

There seems to be an association between deep convolutions and whorls in mammal ears and high frequency sound sensitivity—intersecting with the animals' behavioral adaptations to precise sound localization. It would be interesting to compare the depth of these convolutions in various animals against their reliance on other perceptual cues. I suspect that pronounced convolutions are more evident in animals that depend on sound for close range or finely detailed auditory cues and less evident on animals that rely more on olfactory or visual cues.

For example; humans will locate and orient toward a non-visible sound source, but they will not track it acoustically once it is discovered, rather they fall back into visual cues to substantiate the source of the sound (we'll poke around to find it). Cats on the other hand, will track a mouse scampering in the weeds, or some other hidden sound source with their ears and head, and not depend on visual cues to anchor their tracking.[63]

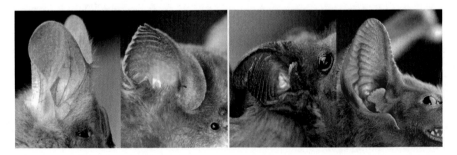

Fig. 4.3 Variations in bat ear morphologies (photo courtesy of Rolf Mueller)

The argument associating pinna convolutions and auditory dominance is particularly persuasive if you look into the world of bats. While the *megachiroptera* such as the flying foxes can see just fine, the *microchiroptera* have poorly developed eyesight and are thus known to be "blind."[64] These microchiroptera are a clear example of an animal that thoroughly depends on sound. This is strongly evident in their appearance, as most of their distinguishing individual features are adaptations to hearing and sound production, with amazing sound projecting noses that look like plants or flowers on their faces and a range or ear shapes that are equally phenomenal. A common ear theme across the species is a delicate folding and ridgework running laterally around the inner surface of their pinnae. These folds are reminiscent of the fine patterns that dance across the liquid surface of musical wineglasses—as if the minute sound waves of their echolocation calls were frozen in ear tissue.

Another characteristic theme in bat ears is a very pronounced, movable "tragus" often sticking up like a dagger, well in front of the pinna. The sharp tragus splits the incoming acoustical signal into two, creating a multi-path signal distinct to each ear. Its mobility allows the bat to fine-tune the signal in the time domain, providing the bat with a broader set of acoustic tools to work its sonar magic around.[65]

In addition to the pronounced tragus, the finely etched textures within the bat ear—like convolutions in the human or cat pinna, help the bat really hear its surroundings, which a bat can do in spades. It is a feat enough to locate, apprehend and eat 600 flying insects per hour (one mosquito every 6 seconds) while hunting in the dark, but some bats (*myotis vivesi* and *Noctilio*) can even "see" fish swimming underwater just by the acoustical signature of ripples left by the fish on the surface of the water. They can perceive these subtle perturbations while flying along the surface of the water, even with their sonar bouncing off and away from them at a low angle of incidence. Somewhere within the delicate little reflecting grooves on their ears, bats can gather, extract and process the fine time-domain information that returns from their sonar as they bounce sounds off of their surroundings—giving them a really complete "picture" of where they are.

Bat echolocation is such an intriguing phenomenon that it comes up in almost every bio-acoustic conversation I have with non-bioacousticians. When people think about animal hearing, bat hearing figures pretty strongly; but a close second is the discussion about elephant ears. Given that the elephant's ears are so grand—particularly the African bull elephant, they really capture the imagination. Elephant's ears are flat and non-convoluted; they hang on the sides of their heads, and they are quite mobile—they can flap.

The elephant's flat ears may have functional commonalities with the "social ears" of primates, matching their sociable demeanor. The elephant is not a predator nor much of a prey animal, so their ear mobility isn't really required for hearing predators or apprehending dinner. The large size of their ears seems to have as much to do with thermal regulation and visual presentation as it does with focusing sound. Elephants will tilt their ears like satellite dishes to pick up distant sounds, but they also flap them when they are hot, or when they are establishing boundaries with humans, lions, or hyenas. Elephant ears are a dominant visual feature of the elephant, but they don't serve as their only sound gathering tool, rather the hearing job is split between their ears and their feet.

If you look at the skeleton of an elephant, each foot is comprised of a delicate set of "toe bones" pointed down toward the earth. In a living elephant, these toes are surrounded by a soft "balloon" of fat and tissue. The softness of their feet is really apparent if you have ever been around strolling elephants—their gently padded footfalls make almost no sound whatsoever. Friends who have slept in the African bush have told me how shocked they were on waking in the morning to find that their tents had been completely surrounded by elephants over night (and nobody got squished). Elephant's feet were not exclusively designed to allow them to sneak around campsites at night, although they do work well in this capacity. From a mechanical standpoint, their feet couple well to the earth, serving as acoustical transducers. Elephants use their feet as both seismic transmitters as well as receivers, allowing them to communicate with each other miles away.[66] This system of bone conduction through their feet could be enhanced by the fatty tissues in their cheeks, much like the lipids in whales and dolphins, coupling the bone-conducted sound into their middle ears—though this speculation has yet to be tested.

Elephants hear and sing through the earth, and the seismic sounds that they hear and transmit are all low frequency, long wavelength sounds—10–20 Hz. Here is another instance where long wavelengths are used for their reach. The low frequency vocalizations of elephants have been long been known to the Tutsi and Hutu natives of central Africa, and incorporated into their songs,[67] but this elephantine characteristic was just recently revealed to science. Somewhat inadvertently biologist Caitlin O'Connell-Rodwell noticed how elephants would respond to seismic events like thunderstorms or "herd culling" at extraordinary distances of 50–150 km. This prompted her to look into the elephant's perception of earthborn sound.

Sensing sound through substrate vibration is actually fairly common in the animal world. Dr. O'Connell-Rodwell's familiarity with an insect substrate sensing behavior had her notice a similar behavior in elephants, cluing her in. In critical listening situations—the advance of an adversary or other communication settings, certain insects would lift some of their legs off of the leaf or plant surface they were on to better couple their sensing legs to the substrate. She noticed that preceding the arrival of kin or some other coincident seismic event, the elephants would freeze and lift one leg up to better settle their contact feet into the ground. She drew the parallels and began researching the seismic communication behavior of elephants.

We humans mostly have our heads up above the ground and are adapted to airborne sound (though there may have been a time prior to pavement and our use of shoes that we were more sensitive to acoustical vibrations in the earth). Nonetheless we still use the metaphor of "keeping your ear to the ground" to indicate a readiness for advance notice of some upcoming event. Of course I have tried this, but since the widespread mechanization of ground transportation and the concrete segmentation of the plains, the earthborn sound-scene is too cluttered to hear any herds of buffalo or distant thunderstorms rolling across the prairie.

Ears to the Ground

When I do put my ear to the ground, part of my inability to hear distant ground waves through the earth is due to the cacophony of large but closer sounds: Freeways, cars and trucks; airplanes overhead, buildings vibrating with air-conditioners and refrigerators. But it is also due to the breaking up of the land by paved surfaces, concrete foundations, buried pipelines, and drainage ditches—all chopping up the surface of the land with impediments to long wavelength ground waves. So the increase in earthborn vibrations and the partitioning of the long range sound channels make this "ear to the ground" phrase a mere metaphor in our times.

It has only been in the last ~150 years, since the arrival of the locomotive that any human generated seismic sounds exceeded the volume of natural sounds. Prior to the mechanization of travel, the hoof beats of migrating beasts, the rumbles of thunder, and earthquakes were about as loud as the earth usually got. In these earlier times the galloping of approaching horses probably had a pretty good seismic reach. But the plains have been fenced and the buffalo have all been slaughtered so there is no longer much concern for reckless wild things thundering into our lives from over the prairie horizons. Though if we scale seismic sounds down a bit into the priority framework of the little animals who reside close to the ground—mice, snakes and lizards, toads, rats, and gophers—things thundering into their lives is a matter of daily occurrence, and the "ears to the ground" phrase is an important reality.

Sound at or near the earth-boundary differs from free-air transmission of sound. Once airborne sound hits the ground it sort of spreads out like spilled milk—adhering to the flat surface and "time-smearing" across frequencies. As a consequence, these small, close-to-the-ground creatures don't tend to have pronounced or detailed external ears. Even the relatively large ears of mice and rats have no extraordinary convolutions or shapes fine-tuning them to specific acoustical details. But the physics of near-boundary sound works in their favor in a few ways. Human ears are well above ground and are usually not very close to reflecting surfaces. If a specific sound occurs, we usually hear it only once as the energy passes by our ears; if this sound reflects back toward us from of a nearby wall or up from the ground, we may hear it twice. We perceive the time difference between the direct and reflected sound as spatial information that gives us localization cues. In the mouse situation, due to the proximity of their ears to the ground, the direct and reflected sounds arrive at approximately the same times. In this setting, the direct and reflected sounds positively sum, doubling the volume, so there is a bit of gain given to them as a result of their being close to the ground.

You can hear this phenomenon for yourself if you care to lie around on the floor with a sound source such as the radio playing. By raising and lowering one of your ears to the floor you will notice changes in amplitude and sound quality of the floor level sound. This works best on hard surfaces, and you can sense it better of you use an earplug in your ceiling-facing ear. (You may want to clue your housemates in before you do this.) Bearing in mind that your ears and head are not really well adapted for this, as you raise and lower your ear to the ground you will nonetheless notice that the near-floor sound scene is different—and louder—than it is higher off the ground.

Another characteristic you might notice when you are down on the floor is the way distant noises sound closer—apparent if something or someone is walking around in your experiment space. This feature increases the apparent reach of hearing for animals living close to the ground.

If we add to this the ground vibration sensing that these animals have available to them through their feet, bellies, genitals, and tails contacting the ground, these ground level animals may have a seismic sense akin to the seismic hearing of elephants, but on a much smaller scale. If these "ground-level" animals do take advantage of these tactile sound channels, it would explain why they may not need remarkable looking outer ears.

Unfortunately most audiograms for rodents take place in wire mesh cages, transparent to sound, and with flooring which is not necessarily set up close enough to a boundary layer to provide this ground wave advantage. Their ears are tested for sound sensitivities—but not their bodies, in settings that assume exclusive air-borne transmission of sounds. (Typically system calibration microphones are placed at the specimen head location at ear level.[68]) This might account for the relatively poor low-frequency performance of these animals in the lab, when in fact they may be quite sensitive to the low frequency energy in the footfalls or wingbeats of incoming predators. The audiograms don't show it, but these ground-level animals are particularly interested in lower frequencies for predator avoidance.

Substrate-borne vibration perception through legs, belly, and body can come into play when the ears are too small to capture and resolve long wavelengths. The earth, twigs, leaves, or branches that small animals reside upon can confer both airborne sound and mechanical vibration into their bodies. It is by way of substrate borne vibration that leaf-miner beetles are alerted to predatory wasps whose wingbeats vibrate the surface of their leafy home. The leaf miner bores tunnels in the soft tissue between a leaf's outer surface membranes. The membranes serve as a sounding board or "drum head" collecting and amplifying the sound of the wasp's maneuvers. When the miner senses the vibration of the hovering wasp it freezes so as not to betray its location. (This acoustical paradigm cuts both ways, because the wasp is also listening for the miner's chewing—amplified by the resonating leaf, enabling it to locate the unwary prey.[69])

Substrate vibration can also serve as a surreptitious communication channel. Small animals that live in wood (like termites) or on branches (like lizards and insects) have ways of generating and sensing substrate-borne vibration. This adaptation helps them keep them in touch with their kin without alerting predators—who may be listening for airborne sound cues about their whereabouts. This form of communication is called "tremulation" and is found in insects,[70] amphibians,[71] and reptiles.[72]

Pursuing a lead on reptile substrate communication, I contacted Kenneth Barnett, who at the time was with N.Y. State Dept. of Conservation. As a lizard enthusiast, Kenneth more-or-less "discovered" reptile substrate communication when he sensed a buzz through his bones as he was handling a pet chameleon (a male *Chamaeleo calyptratus*). He investigated this in some detail with an accelerometer attached to a branch that his chameleon perched on and found that his calls were stimulated by the presence of other chameleons (mostly girls). The pitches were in the range of 90–150 Hz, but he played me some higher owl-like hoots he recorded that were more in the range of 300 Hz. (These hoots were somewhat eerie, sounding like "ghosts"—which might account for the Yemen and Madagascar folk tales that feature the chameleon singing creation songs into the forest.)

Kenneth didn't explore the mechanics of this in detail as he is not a physicist, but if you examine the chameleon form, it yields some clues about their tremulation mechanism. Their feet are formed to firmly grasp small branches, with toes that wrap around a branch and really lock into place. This provides for a good coupling both as a transmitter and receiver of the tremulations. Many chameleons also have a tall bony crest, or "casque" on the top of their heads, making them appear rather dinosaur-like. This crest has some mass to it and may serve as an inertial mass or "counter balance" from which to overcome the mass of the branches they sing through while imparting their love songs into the bush.

It turns out that tremulation is fairly common adaptation and is imparted into a substrate by any number of methods. My take on the chameleon casque is somewhat speculative, as I don't know if anyone has examined this in detail. But if the crest is used for inertial mass, it could be a common adaptation found in other lizards and insects also known for tremulations such as the thornbug treehopper (*Membracidae*) whose crest makes them look a bit like rose thorns. Other insects that don't have the crest will perform rapid "push-ups," using their whole body mass to vibrate their perch. Some katydids tremulate in this manner even, while they also "stridulate" (the term for sound production through the rubbing or flexing of body parts). Their dual-mode communication channels are a likely adaptation to one of their key predators, the "gleaning bats." The katydid's periodic airborne advertising calls cautiously bring in their kin (without revealing their location to the bats), and their tremulations maintain contact with their kin once they are situated within tremulation-sensing distance (greater than 10 ft in one citation).[73] The katydid's substrate sensing is done through mechanoreceptors in their legs as they grasp the branch or surface they're on, allowing them to sense substrate displacement differences in their legs against the inertia of their body.[74]

The real "ear to the ground" animals are the subterranean inhabitants like moles, gophers, and mole rats. There is little call for vision in their world (mole rats are even visually blind). But there is ample use for sound communication. Typically their audiograms have been performed in cages, showing a limited hearing range, but a recent study of mole rats has determined that these creatures use "seismic echolocation" to expertly navigate around obstacles in the earth. They create the seismic signals by drumming their heads against the top of their tunnels. They can't easily pick up these vibrations through their ears while drumming, so they rely on mechanoreceptors in their paws to hear the seismic signal return.[75] This low frequency sensitivity would also allow them to hear the footfalls or predators, the drumming noises of kin, and the slithering friction of snakes—sounds that they need to be alert to.

It is evident that outside of our airborne sound channels there is a whole lot of shakin' going on. The earth, the plants and trees, leaves and logs—even the timbers in our houses are all channels for sound and vibration used by the creatures that dwell in these environments; animating them with their activities and "song," and listening in with all manner of sensory systems.

(Some) Bug Ears

The nervous system of vertebrates are armored and protected by the hard bones of our internal skeletons. Our spine encases our spinal cord, and we have all evolved to have our brains encased within a protective envelope in our heads. This has allowed for the centralization of many of our senses—with the ears, eyes, nose, and tongue all in close proximity to the brain.

Arthropods—our joint legged friends (including the insects, spiders, and crustaceans), have exoskeletons protecting all of their internal soft tissues, so the distribution of their nervous system is not so constrained.[76] The advantage here is that the senses (and their associated neural processing) can show up on their bodies where they are needed—and not necessarily in their heads. So what we commonly call "ears" appear in various locations and in a myriad of forms, structurally adapted to their habitat and behavior. Sensing hairs, diaphragms, slits, tensor muscles, antennae, cavities, resonators, statocysts, and sacs occur in various places all over the bodies of these animals, binding them into their specific acoustic environments.

Arthropods are the most widespread group of animals. The diversity and breadth of their niches are uniquely expressed in their sensory adaptations. Dwelling on the deepest sea bottoms, and blown by the winds way up above the clouds; from eyelash mites to water whirligigs, stone crabs to ghost shrimp, fig wasps to fruit flies, centipedes to wolf spiders—these creatures all have sensory adaptations to their specific worlds helping them feed, breed, avoid predation, find prey, navigate, avoid bad weather, or lock into their community. Their sensory systems can be adapted to the one specific quality of their habitat that gives them the survival edge in their niche. With phonoreceptors specifically tailored to sense exclusively close range, or specifically long range sound; they can be finely tuned to distinguish particle motion and/or pressure gradient energy, or tailored to perceive delicate coherent sounds imbedded in high density noise. In this phylum we find focused perceptual adaptations to substrate vibration, airborne sound, phase coherency, short impulses, barometric changes, and even pitch discrimination. I find it amusing that with the exception of vibration-sensing antennae, none of the arthropod sound receptors show up on their heads. Typically they are found in or on the legs, or sometimes in or along the sides their bodies. They can show up in symmetrical orientation—in pairs for "stereo localization" but they can just as readily show up in sets of six or eight, as single organs, or as sensing areas that don't necessarily lend to clear stereo symmetry.

For most people the discussion of insect sounds begins and ends with crickets and cicadas. But if you are out in the country, or even in a suburban area, it's worth going outside for a moment to give the air a listen. If the wind is still and the sound-scape is not cluttered with traffic, chainsaws, or overhead jets, you might notice that the background noise is comprised of the buzzing hum of other insects.

We take for granted that the buzzing of flies and mosquitoes is just a product of their means of locomotion, but their wing beats are also a signature sound of their species. This was first noticed for the record in 1878, when a row of electric lights in a New York hotel attracted male mosquitoes who mistook the buzzing frequency of the lights for the sound of breeding females.[77] So while it may seem that the high pitched buzzing of the mosquito was designed to give humans an auditory preview of Hell, it actually serves a slightly higher purpose of gender identification for the mosquito.[78] Even in the ~2,000 species of fruit flies (*Drosophilae*)—whose genetic-lab personas would indicate anything but partner prejudice, avoid cross breeding by listening to the specific sounds of their prospective partner's wing beats. If the pitch doesn't match their particular species, they move on.[79] (Both the mosquito and the fruit fly can discriminate pitch, but unlike the pitch discriminating mammals that rely on a cochlea, both of these animals sense pitch with their antennae,[80] which look much like plume feathers.)

For me one of the most impressive qualities of arthropod hearing systems are the diverse manners in which such small animals accurately localize sound sources from acoustic wavelengths that are much larger than their own bodies. With wavelengths larger than their bodies, they can't rely on the inter-aural time and intensity differences that we humans rely on. They don't have pinnae to create secondary reflections, and in many cases the sound sources they seek are deliberately obscure or complex.

The branch-perching insects mentioned above can at least assume that the sound source is either up-branch or down-branch from them, but the parasitic fly *Ormia ochracea* can localize the phase-ambiguous song of their cricket prey with an accuracy of two degrees by using a set of mechanically coupled membranes at the base of their neck. These membranes are less than 1 mm apart and mechanically coupled by a cuticle. This setup is less a stereo pair of ears than a set of differential diaphragms shadowed by the chin of the fly.[81]

Most animals use time domain and phase information to localize sound, but due to the proximity of the Ormia hearing diaphragms, they are specifically *not* sensitive to time domain information. This comes in handy for them when locating their cricket hosts. Crickets have a very complex time domain song specifically designed to confound their larger predators. But due to the Ormia hearing mechanism, it is not baffled by the cricket's main acoustical defense; rather its hearing is specifically finely adapted to sense amplitude only.

As indicated above, any animal that calls out to contact their kin runs the risk of alerting predators to their whereabouts. The tremulating animals overcome this by using surreptitious substrate sound channels, but animals that are not dwelling on the same substrate don't have this advantage. One strategy that community animals use is "chorusing." The idea being that if all animals are calling at the same time, a predator will have a hard time finding any individual. This strategy is akin to fish schooling; a predator is faced with too many choices and is unable to select the most likely prey as the fishes dart away in evasive action. But chorusing insects—crickets and cicadas—are not constantly moving like schooling fish, so any sound that an individual makes can easily betray their location to a patient predator perching nearby.

Field Crickets (*Gryllus* sp.) have an interesting solution to this difficult conundrum. If you have ever tried to catch a cricket, you may have noticed that they are hard to locate by their sound alone. Their song presents and ambiguous soundfield, so when you are close enough to pinpoint the sound source, it seems as if the sound is coming from an area significantly larger than the cricket. This is called "animal ventriloquism," which is a strategy that various insects and birds use to conceal their whereabouts. Often it involves altering the time-domain component of their sound to create a phase-complex signal. This confounds predators that use time-domain information to locate their prey. There are a few interesting ways that animals create ambiguous soundfields; some involve synchronization with the sound-body of the community, others involve the mechanics of individual sound production. Crickets are sound-community animals but individually territorial, so they use a bit of both. I've been fortunate to have had the opportunity to closely examine the mechanical side of this in the privacy of my own home.

Living in the country, I have had a number of crickets as house mates. Crickets seem to have a talent for visual recognition and memory, so once a cricket has identified that I am not a predatory threat it will continue to sing even if I am just inches away. Most people are familiar with the way crickets stridulate by rubbing the edges of their hard "elytra" or "forewings" together with a file edge on one wing and a plectrum edge on the other. The elytra are furnished with tuned resonating surfaces called "harps" which account for the bell-like pitch of the sound. This sound generator is enhanced by another feature of the cricket's song. In my private audience I noticed that these cricket's have soft, crumpled-looking "alae" or "under wings" beneath their elytra, which they open and extend as they sing. These irregularly faceted under-wings probably resonate a bit while reflecting the sound produced by the fluttering elytra.

I suspect that the sound of the elytra reflects off of, and resonates with these irregular under-wing facets, randomizing the signal and creating a complex sound signature with extreme phase or "time domain" information in it. This would account for why the cricket's sound seems both larger and less tangible than what

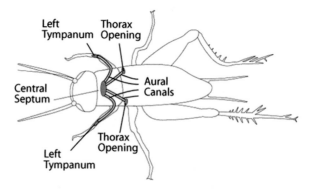

Fig. 4.4 Cricket Auditory canals. Some cricket hearing systems integrate relative amplitudes of signals received from four inputs to determine proximity and direction of the signal source

we might expect from this small creature. You can get a similar sense of this effect by whistling through a spinning fan; the blades bounce and diffract your whistling around in a pseudo-random manner creating an ambiguous, cricket-like soundfield.

The ear of the Ormia fly explained above is not sensitive to phase information, only to amplitude information, so they are not fooled by this time-domain ruse. Of course the crickets need to overcome this confounding ruse as well if they are going to locate their sweethearts within the ambiguous soundfield of their community. They handle it in different but equally ingenious manner. Crickets have tympanic membranes on their front legs, as well as openings on the sides of their thorax. These acoustical pathways all commute to a central membrane in their thorax. This allows them to compare the relative amplitudes of all four inputs, two from front leg tympanum and two from the thorax openings. The central membrane converts the amplitude signal into a relative phase signal, which is mechanically tuned to the characteristic frequency of their species' song.[82]

Crickets have a tuned sensitivity to their prospective mates in the 3–9 kHz band, but many also have a secondary sensitivity to ultrasonic sounds above 15 kHz, which they use to alert them to predatory bats.[83] They also seem to have some manner of sensing the low frequency signature of approaching predators, accounting for why they usually shut up as we get near them. With this, the cricket's acoustic sensitivities show up as three separate perceptual modes that don't really overlap: bat alert, terrestrial predator alert, and love songs. This tells us something about their real life worries and cares.

Like the crickets, many nocturnal moths are also one of the dining delights of bats, so they have also adapted to hear ultrasonic sounds. While there is no evidence that these moths use ultrasonic sound to communicate to their kin, they have co-evolved to hear the bat's echolocation signals by way of tympanal organs in their thorax.[84] When they hear the calls of incoming bats, they will fold their wings and drop like a stone in an evasive maneuver. In the case of tiger moths (Fam. *Arctiidae*)—and underwing moths (*Catocala relicta*), and perhaps other unstudied moths, this evasive action is accompanied by a loud ultrasonic "scream" that they emit in the critical moments just prior to the bat's hoped-for interception with the moth.[85] The moth's scream could be a phase complex wave front, creating a "ghost" acoustic image of their body like a "jamming signal" to confuse the bat, or it could just be like screaming "Boo!" startling the bat into losing its focus at the last moment.

Most insects are niche specific and don't need to make complex decisions. As a result, insect behavioral studies can key into a specific stimuli, eliciting predictable responses. These animals are not making "decisions," rather their behaviors seem more genetically programmed. This is one reason why insects make good behavioral study subjects; when a specific behavioral stimulus is introduced, the animal reliably responds. Web building spiders on the other hand are not so predictable. They seem to be all about decisions, starting from where to put the web, to how to respond when something trips it.

The webs that spiders create are actually sensory extensions of their active space. Their survival success hinges on where they place it and how well they manage it.

While spiders have multiple eyes to cue them into their immediate surroundings, their eyes are mostly used in close proximity around the action scenes of predator/prey conflict.[86] Vibration sense is what gets the spider to this point, with three different vibration sensor types in each of their eight legs responding to the various motions of their webs.[87] These help them sort out and localize predators and prey, and also help them negotiate their tenuous dance of love.

Spiders live in a warped sense of time and much of their life consists of interminable stretches of just sitting around in metabolic suspension. This is interspersed with short, dense episodes of rapid decision making when something strikes their web. Not everything that strikes the web pleases the spider. In my area there are tiny little dust gnats that swarm in the late afternoons of spring. They get stuck in the webs and are mostly wings and frith—hardly enough to get a fang around. When they get stuck, they just hang there like festoons and can really bog a web down. Removing them one at a time as they occur expends too much energy, so a spider will wait until there are a dozen or so (or when the swarm is over), then using a type of vibration-echolocation to find them,[88] will go out to cut them loose and tidy up the mess.

When a large moth gets snagged it is a different story. A frantic moth trying to escape a web can wreck havoc on it, tearing it asunder to the degree that the biological expense of rebuilding the web eats into the value of the meal. In this situation the spider needs to rush in and subdue the moth, tacking additional silk tethers on its wings as fast as possible. Smaller moths or lacewings are much less trouble for a satisfying meal, and the "subdue" phase is much less urgent.

The real thrill happens when a wasp gets tangled in the web. Wasps and spiders are mortal enemies—much like Godzilla and Rodan, and some of the most dramatic interactions in nature occur around their relationships. Wasps typically get stuck in spider webs while looking for a spider meal, so when they get tangled, the event can unfold as a "eat or be eaten" affair. The spider has to somehow subdue the wasp while avoiding its grasp and sting.

Spiders gauge the scope of these various interactions—the gnat, the moth, the lacewing, and the wasp—by the vibrations that they produce in their webs. They can localize the interloper within a few degrees and a few body lengths from their perch by the vibrations sent out along the threads. They can then go out and deal with the conflict as required.

I lived out in the forest for a while in a house with a wrap-around verandah. Due to the forest setting there were so many spiders around my home that it seemed like I was more in their habitat than they were in mine, so I had to choose my battles regarding their domination of habitable space. In this house I let them take over the eaves of the verandah.

One afternoon I noticed wasps swarming under two of the eave sections, so I went over to investigate. This turned out to be quite a drama as the wasps systematically approached the webs and ever-so-gently tickled the threads with their long back legs. This vibration would bring the spiders out—probably expecting a lacewing or some other easy bite, at which point the wasps would grab the spiders and trundle them out into the forest back to their own nest. This went on for an hour or so, the wasps systematically gleaning one section of eaves, then moving on to the next, all around the perimeter of the house, taking most of the spiders away with them.

A few days later I was watching one of the remaining spiders in her web as a wasp approached. When the beating of the wasp's wings came close enough to excite the web, the spider quickly retreated into a protected corner. It is possible that this spider was a survivor who associated the vibration of wing beats with predation.

Arthropod sound and vibration sense is a broad and diverse field, well beyond the few cases explored above (and well beyond the scope of this book). Other terrestrial arthropods, as well as their marine arthropod cousins—shrimp, barnacles, lobsters, crabs, mites and such, have all tuned into their own habitats with their specific relationships to sound and vibration—each with unique sets of environmental constraints and conditions. If you can say one blanket thing about arthropod hearing, it is that they are sound specializers, with a fabulous array of mechanisms to acutely tap into their surroundings. This provides a whole new dimension to the phrase "as acute as a bug's ear."

Ears Take Flight: Bird Ears

Across the panoply of acoustic adaptations in the animal kingdom, a few have really stimulated imagination of scientists—who bring them in for closer scrutiny. Bats and dolphins share that distinction, largely due to the magic of echolocation. But close contenders in the "most research done" category are on the acoustic adaptations of owls. Owls are nocturnal and known by the large light collecting eyes that enable them to see in their dark domain. But the owl, with its adaptations to night sounds can be also considered a flying ear. On their broad silent wings, owls can locate their night-time prey hiding in tall grasses or under a blanket of snow—silently descend, and swiftly kill an animal before the victim's terror even arises.

Their ability to precisely locate concealed prey relies on a number of fabulous features and qualities of their hearing system. In the extensive literature on owl ears, probably the most commonly mentioned features are the dish-shaped feather ruffs that form the owls face (serving as a sound collector) and the asymmetrical displacement of the left and right ear. On a horizontal plane, the left ear is located about a quarter to a half inch higher than the right ear, giving the owl the ability to acoustically triangulate on a sound source.[89]

One early morning, just as I was embarking on this chapter, I came across a barn owl (*Tyto alba*) that had met an untimely end as road kill. Of all owls, these fellows are probably the most ubiquitous, being found on all seven continents, so I couldn't have asked for a better research partner. Holding his light, fluffy body in my hands, I could easily see how the soft feathers on his wings and body attenuate the sound of wind turbulence. The facial ruff is not "dish-shaped" so much as it is the shape, and almost the same size of two human ears connected together along the broad

Fig. 4.5 (a) Barn Owl. Stiff facial feathers serve as a sound collector akin to the function and shape of a human auricle (photo: Johan Doe). (b) Dead Owl. Facial feathers pulled forward to reveal the auditory meatus (ear canal) (photo by author)

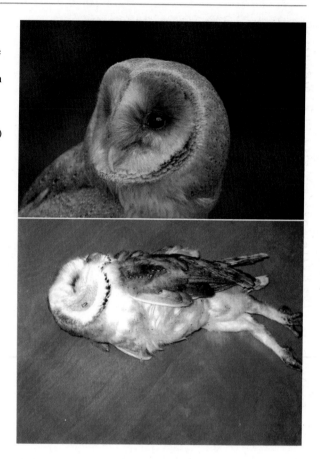

edge. Each side of this heart-shaped form traces the same "phi curve" of human ears previously discussed, even following the same earlobe curves at the bottom.

The ruff is a thin layer of stiffer feathers that stand off from the face. You can place your little finger behind this layer and bend it forward to reveal a surface of bare skin and the auditory canals leading directly into the skull. The texture and shape of the ear from this view is particularly human-like, with convolutions around the outer ear canal very similar to our own. If there is a distinct difference from our human ears (aside from the feathers), it is that the owl has a "pre-aural flap"—like a tragus, extending out away from their eyes and shading the ear hole from direct sound. This feature is about ½" square, wafer-thin. So like the bats, this pronounced "tragus" splits the sound path in two at each ear, providing a time domain tool for localizing sound sources. This tragus articulates, so it would be fascinating to know how they mediate its movement as they focus in on the sounds they need to hear. Given the clear specializations of the owl ear, their ability to localize hidden sounds does not surprise me, although if I were a mouse blithely foraging under a blanket of snow I could be surprised—and the surprise would be swift and cruel.

Owls are a peculiar case in the bird world, with outer ears that seem more akin to primate ears. Other bird ears reflect an auditory pedigree closer to reptiles,

Fig. 4.6 Inter-aural canal in reptiles and birds help localize sound sources by comparing relative pressures between the two tympanic membranes. (This echoes the common ancestry in these two families)

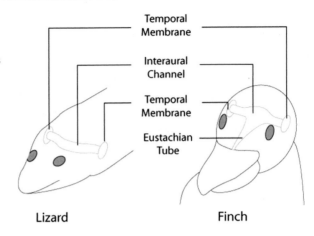

Temporal Membrane

Interaural Channel

Temporal Membrane

Eustachian Tube

Lizard   Finch

mirroring our understanding that birds and reptiles hail back to a common dinosaur ancestry. One similarity between the reptiles and the birds is the way that they localize sound sources. Both of these families lack an outer ear or any sound collecting structures like our pinnae to help them localize sound. Because their ears are fairly close together, interaural time and spectral differences are not distinct enough to provide good localization cues. What they do have is an interaural canal between their ears through the cranium, so that sound impinging on the outer surface of the tympanic membrane on one side of their head affects the inner surface of the membrane on the other side—in the opposite direction—by way of the interaural canal. This serves as a differential pressure detector, accentuating the sound differences between the ears; so sound arriving from one side of their head effectively cancels the same sound arriving at the other side, helping birds and reptiles distinguish the direction of a sound source.[90]

There is no compelling environmental reason for lizards not to have outer ears, but there is a good reason that birds don't. Floppy head-mounted appendages would be an aerodynamic disadvantage for birds. Fortunately for the birds (with the exception of the owls), they don't need to accurately pinpoint sound sources. Given that bird songs mostly serve as territorial announcements and songs of seduction, they are typically stationary when singing, perched on some branched beacon or landmark, so in locating kin, they just need to determine proximity and basic direction. In predator evasion they only need to be alert to the threat's approach direction so they can determine which way to take flight. These characteristics support the argument that acute localization is not critical for birds.

You might expect from their species-specific vocalizations that birds don't need a wide range of frequency sensitivities; they just need to hear the songs of their kin. But some birds have a remarkable sensitivity to infrasonic energy. In homing pigeons this sensitivity has been measured down to 0.05 Hz (one cycle over 20 s).[91] With this sensitivity to slow dynamic pressure gradients, it is likely that these birds can sense barometric pressure changes and perceive trends of turbulence along their flight path. Hearing perception in this realm would allow a flock of migrating birds to determine the scope and breadth of an oncoming storm—days before its arrival.[92] It would allow them to know if a storm is a result of the seasonal instability of early autumn, or if it is the true onset of winter. "Hearing" the weather might cue them in to the safest opportunity to begin their seasonal migrations—coordinated to best take advantage of smooth tail winds and stable pressure fronts.

There is some informed speculation that the ultra low frequency sensitivity of migrating birds helps them navigate, and that they have mapped the terrain of their migrations by knowing the sounds of the earth as it is played upon by the weather— listening to how the winds play over mountain ridges like huge flutes and how the waves play the shores in their deep rhythms, slowly modulating the pressures of the earth's atmosphere.[93]

Many songbirds migrate nocturnally, when visual cues are not pronounced. And while many migrating birds don't travel in conspecific flocks, they move by circumstance in groups with other birds. These birds keep tabs on each other by the use of "call notes" or "chip notes"—brief announcements of their presence to other birds in their migrating flocks. We usually miss these sounds as they often occur thousands of feet up in the air, though they can sometimes be heard in the early hours of dawn as the birds return to earth for the day. These chip notes are a part of bird vocabulary that serve as "placement tools"—a set of sounds that enable birds to establish their presence with each other on the ground and to avoid collision in flight.

One July a few years ago was out at a nearby lagoon watching the seabirds work their nesting grounds. It was a hot, pregnant day with countless species busy securing their nesting areas and gathering food for their offspring. Within this horizon I was transfixed by a tight cluster of hundreds of plovers in flight rapidly ducking and weaving across the lagoon, their white breasts flashing together as they caught the reflected light of the summer sun. The sound of their collective wing beats was a dense flutter that I can only describe as "butterfly thunder."

It was only at the apex of their dips, when they all silenced their wings for a brief moment, that they exclaimed with a volley of chip notes; returning immediately to powered flight. They each knew where they were in their fast formation, but something larger was binding them together. Since that time I can no longer look at a flock of birds without seeing a single body.

It could well be that like the "proximity sense" of schooling fish, the synchronized ballet of flocking birds is bound together by the low-frequency pressure oscillations that they collectively generate—responding to the wingbeats of the flock over the theme of air turbulence within which they are flying.

In support of this conjecture, I was just last week out at the same lagoon—now being September, at the end of the breeding and nesting season and in the beginnings of fall migration. The birds that have not yet left on their migrations are no longer driven by sex and procreation; rather they are awaiting Nature's signal that it is time to depart. In this season of migration, there is no longer the tight groupings of species and kin; the shorebirds congregate and feed with other shorebirds—the curlews, plovers, sandpipers and godwits all wading in the shallows together and probing the mud for food. Occasionally something would startle the congregation and they would all rise and take flight. From the smallest plover to the largest dowitcher, they would synchronize their darting and weaving; and while their diverse flock was less urgent than those flocks of the breeding season, their white breasts and underwings would all flash simultaneously as they do when tightly grouped with their own kin.

All of the animals discussed above have evolved acoustical energy senses tailored to their specific connections to their environment and role in their niche. The physics and mechanics of their adaptations express the multitude of ways acoustical energy impinges on our surroundings—surroundings that are all resonating with the energy of movement and the vibrations of life. The seismic shifts of the earth, the pressure gradients of the weather, the rustle of leaves, the pulsing of adjacent bodies, the fluttering of wings, buzz of a branch, or a knock on the door. These sounds inform us about our environment and tell us something about others who share it with us.

♦

For years after the body of a young gray whale had washed up and decomposed on the beach I would occasionally go out and visit her bones. Time buried her in the sand. The sedge and verbena sprouted over her resting place, creating a comfortable mound to rest against.

It was only about a year after her demise that I took a friend out to witness this little altar that the elements were crafting. We were sitting in the sand facing inland for a while, looking over the exposed bones. I wanted to move on, so I casually commented that we should set out and look for some living whales.

We stood and turned around to face the sea to find that two Gray Whales, a mother and calf, were gamboling in the surf directly out from the grave—and no more than a few yards from the shore.

We all remained in each other's company for an hour or so, paying common tribute to a very special life that had passed, and like the verbena and sedge to honor the gift of which we are all a part.

Communication: Sound into Form

5

> *Communication...is now recognized as the mechanism by which all the essential interactions between organisms are accomplished; a system of transmitters and receivers which integrates organisms and coordinates their activities into functioning social groups or communities much as the nervous system coordinates the activities of the tissues in a smoothly functioning organism.*
>
> (John T. Emlen, Jr. in his introduction to *"Animal Sounds and Communication"*)[1]

I have returned to the Sierra hot springs pool that opened this book, nestled in the skirts of an alpine meadow. It is very early this spring morning, the sky still dark, though fading to light behind me. A forest of cedar and Douglas fir slopes up to the jagged ridgeline; before me lays a grassy marsh. To the east the horizon begins to show the promise of a new day, dimming the stars of Cassiopeia—hung upside down there by the Gods for her hubris. The Big Dipper lays upright across the north sky; stable and open, ready to receive. Bats tangle above the pool, polishing off the last of the night's mosquitoes before turning in. My soundscape is intimate; the drip of water, the sound of breath.

Suddenly the hoarse single chirrup of a robin shoots through the trees. As if in alert response, a quail's rhythmic wallow reports from the meadow. Then almost at once, birds of all stripes and feathers rise up to sing—chirb, twitter, buzz, click; low and bay, gargle and pliew, fling and vleet. It is loud, urgent. A cacophony clutters the soundscape that only moments before seemed empty enough to contain the universe. The dawn chorus has begun.

For the first hour the birds are singing as if it were their last song, without regard for space or cadence; whistles and cliews, shrills and clacks all jumble and clash into each other. But as the last stars fade in the west and the sky takes on a more even hue, the racket has calmed a bit—in volume, if not in density. Somehow the chorus begins

M. Stocker, *Hear Where We Are: Sound, Ecology, and Sense of Place*, DOI 10.1007/978-1-4614-7285-8_5, © Michael Stocker 2013

to breathe in a more common rhythm. By the time the sunlight starts grazing the uppermost mountain tops it feels as if the birds have synchronized somewhat, waiting for clear spaces within which to place their song. This invites sets of "call and response" among the birds: a sparrow, a wren, a robin; a sparrow, a wren, a robin—grazed across the top by the laughter of a troop of woodpeckers or the bouncing zips of a killdeer. Underneath, the rhythmic swells of the quail in the meadow. A cadence is set up—the birds are listening to each other. The fabric of sound has the delicious musical tension of a tight ensemble of jazz improvisers; dropping that exquisite phrase in at the perfect moment, bouncing lightly off the ride cymbal; pizzicato bass stabilizes the free flight. These birds are swingin'! Or is it just my swingin' mind putting order and cadence onto the chaos of mindless birds? (I seem to recall that it was a question like this put Cassiopeia in her inverted predicament.)

I listen for hours: The players change, the shape of the soundscape shifts in size, proximity and pitch. Some birds are in flight as they sing; others bounce from limb to limb, and scuffle with their kin in nearby trees. Some strafe the warm pool I'm in, picking off insects in flight. The overall density of the song diminishes. An occasional thrush lets forth a spectacular sonic diadem; a song sparrow captures a long minute to himself. By the time the sun has bathed the meadow the dawn chorus has ended, only individual songs patch out of the soundscape, now washed by the morning breeze through the pines; the distant sound of a tractor; the scolding of a gray squirrel; the sharp snaps of cicadas in the surrounding treetops promising a hot day.

Not Language

To communicate: The etymology intersects with and brings together words of joining, solidarity, and inclusion. Community, unity, "co-" with Latin roots in gifts and service *munus, munerare*—to give.[2] Communication could be considered a mutual "gifting" implying an engagement, a participation, and an acknowledgement of some form of parity; I communicate with others because including them makes a difference to me. A conveyance of ideas is implied, but the simplest expression serves as a placeholder; a sidelong glance, a shifting of the body, an audible exhale. With communication we are mutually assured by a gift of belonging.

Sound is not germane to communication. All living organisms communicate volumes without the use of sound, but sound is helpful. It implies a willful inclusion, a reaching out to notify others of participation and intention; an acknowledgement that others exist and are worthy listeners.

By exploring sound and communication, it is important to clarify that communication is not language, at least in the way that I understand "language" per se. From its Latin origin *lingua*, language is a subset of communication that is formed by the tongue. It has come to mean a set of rules that put order into the way we convey information to others; rules that we can refer to.

Language is a handy communication tool—a tool that uses sound in a myriad of intricate ways, and a fascinating and fruitful topic; rich with ideas about meaning, form, sound, and purpose. Pursuing these ideas has spawned a systematic study of linguistics—with many brilliant modern practitioners. Morris Halle, George Lakoff, and Noam Chomsky immediately come to mind. But these modern theorists follow a long path of systematic inquiry into linguistics along which one of the earliest Western records is found in Plato's *"Cratylus"* wherein Socrates, Hemogenes, and Cratylus discuss the sound of language elements and the underlying meaning that these sounds impart into words and names.[3]

The origin of language is a question that has titillated thinkers for ages, because the answers to this question would shed light on the origin of our own humanity. On this account it is not surprising that the mythological origins of language often involve Trickster. In *"Cratylus"* Plato supposes that Hermes invented language so that he could prevaricate and tell lies.[4] In Native American tales, Coyote or Raven give the "Two-leggers" a tool which is useful enough to confound them. It is the power of this tool that inevitably leads to the demise of its usefulness—as in the story of the Tower of Babel. These stories are self-referential by nature; without a functional language, the tales could not be told. And while the tales tell us something about language, they also tell us something about ourselves.

It was once thought (and not so long ago) that language was the distinguishing feature of humans—our gift from God that came with the naming of, and thus the dominion over the animals. It was believed that the human capacity for self reflection and our understanding of consequence provided a unique perceptual platform that liberated us from the behavioral laws of "instinct." Unlike the animals, "... humans had tools, words, and reason..." Broad, self-assured assumptions such as these can be both difficult and embarrassing to retract. Similar to the long held assumption that tool usage was the exclusive legacy of humans, it took the humility of an exceptional animal behaviorist (Jane Goodall) to recognize that humans hadn't cornered the tool-use market. She noticed that chimpanzees could speculate about a problem and craft (or engineer) a unique, unlearned solution. By revealing this observation, she took an important first step; others soon followed by sharing their observations. They began to reveal that other animals thought about consequence and time; that they had humor and semantics, that they deliberated and deceived, and that they had an economy of actions—trading favors for allegiance. In this flood of revelation, the determining features that defined "language" as a unique human trait began to erode.

In the early 1960s, Linguist Charles F. Hockett derived a list of 16 design features that constitute human language, including features such as learnability, tradition, system self-reflection, and the ability to generate nonsense or lies. In 1963 these design features were laid out in the context of animal communication by zoologist W.H. Thorpe to evaluate the intersection of linguistic design features and animal behavior.[5] From his evaluation it seems that the only language design feature not found among non human animals is the ability to talk about communication—i.e., to speculate about the origin of language.

So it seems that with our slight digression into scientific methodology, we arrive at a point where the folk tales of the First Peoples may have some factual merit; that the deceptive wiles of coyote and raven, the nobility of the bear clan and the perseverance of the fish nation probably originated in empirical observations of their behavior, which in turn conveyed something to us about communication.[6]

Thus communication is not about language; rather it includes other behaviors as well. Taking this broader view, I feel that it is safe to state that communication is that which binds us to our surroundings; it is the manner of mediating our relationships with those things which we share mutual influence. By not nailing down absolute rules about symbols and syntax, conjugations or grammar, I don't have to exclude my neighbor serenading his roses, or my Auntie talking to her dog from the rubric of "communication." When I hear the ensemble of field crickets and tree frogs pulsing out the temperature of a balmy summer night, I don't need to ask if they recognize each other. And when a squadron of pelicans descends in perfect formation down the steep wind-rise off a coastal bluff, I don't need to ask if they know where they are.

Seeds of Sound Communication

As someone deeply saturated in literate culture, it is difficult for me to shake the idea that language bears discrete meaning; that words are like screwdrivers and hammers held together in the toolbox of grammar and syntax. Within this practical idea of a "language of definitions," it is equally difficult to translate the sounds that constitute the elements of language into the characteristics that they represent. In a literate or written language, words become placeholders; representations of "stuff" which then can be moved around in time and space. They can be hidden from view, saved for posterity, reassembled and quoted out of context, even shredded or burned to erase it from our memory. This literate language has many burdens and responsibilities. Ever up for the task, words and phrases of this language end up generating more words and phrases to clarify what they originally tried to mean (as Lao Tsu explains; "the named is the mother of ten thousand things").[7] Our Indo-European language construction is very effective, but it is so "tool" focused that it's sometimes hard to imagine that there is any other way to communicate; "…these words mean these things, and that's that…" We assemble these word-tools for the purpose of conveying specific meanings.

Regardless of how the words and the meanings are assembled, the material of these words is sound. We form our sounds into vehicles of intention and do it in many ways, depending on our relationships to what we are trying to express and to whom we want to reach. The words we form often come directly from associated sounds. They can include the onomatopoetic sounds told to us by their expresser, as in "killdeer," "patch," or "buzz saw," as well as from imaginal sounds that just seem to match the action—such as "thrash," "bang" or "slice," or the fundamental sounds of reaction: "Ouch!", "Ooh!", and "Ahh!"[8]

Because these words are strongly associated with emotional expressions rather than assigned descriptions, there are those who have taken up the idea that sound communication arose out of songs. Rousseau proposed this,[9] and it continues to be a popular idea. There are musical artifacts of this in our own expression—particularly when we are expressing our feelings—those times when words fail and sound picks up the duty, as in sweet-talking, sing-song beckoning, yelling, bawling, or whining. This "non-symbolic" sound communication is still alive in our own language, though it is particularly evident in certain living indigenous languages. One example of this is from the Kaluli of Papua, New Guinea, who use two forms of language. They use a talking form of communication to get functional things accomplished such as "give me the bag," or "the fruit is here." But when they need to express themselves, they sing a "bird language." This expressive bird language would be used to inquire "what do you think about that girl?", or express "I had a heck of a night last night" and "I love you."

Ethnomusicologist Steven Feld described the Kaluli communication paradigm succinctly: "Talk gets you what you want, need, or feel owed. Song is different because it is a communication from the perspective of a person in the form of a bird."[10] It is their "bird language" that weaves the fabric of their family and culture together—something far more binding and tangible than "having a bag" or "locating some fruit." For the Kaluli, this language affirms the legacy of where they live, in a continuous affirmation of engagement with their surroundings and all who dwell in it—particularly their avian and human relatives. From the perspective of the Kaluli, the language of bird song is much larger than our academic framing of bird vocalizations. The birds are not "vocalizing instinctual behaviors," rather each bird is a character engaged in an ongoing dialog with the forest in their own particular way. Each of these bird characters informs the Kaluli about relationships in the context of their environment or "mutual habitat."

This concept was brought together for me this morning when I was out in the yard watching a chickadee foraging for spiders and mites among the low branches of a fir tree. He was chattering to himself as he hopped, flitted and hung off of the branches, saying nothing in particular. His vocalizations were random, non-repeating, twitters, plicks, and chirrups. It was clear that his singing wasn't territorial, as he came and went without any fanfare. It was also clear that he wasn't vocalizing for the attraction or benefit of a mate, as no other chickadee was within earshot. Unless he thought I was a worthy listener, he was probably only vocalizing for his own benefit.

I don't really know what to make of this behavior in the context of a behavioral science that frames vocalization in economic terms—as a biological expense. This little guy wasn't trying to confuse or distract predators, he wasn't trying to deceive or frighten his prey, he wasn't even trying to attract a mate; he was simply singing while he worked. Could I surmise that he was doing what we all do on occasion; that he was talking to himself? Was he just trying to "keep a groove going" with his enterprise? Was he "singing his surroundings" in his own version of aerial songlines? These are things I will never know, but it was fun listening to him.

"Living sounds" like the chatter of the chickadee, or the idle songs that my neighbor sings to his roses, or the chatty-scolding of my Auntie's interspecies communication express experience in ways that can't be summed up in simple words. When I hear them, they germinate in my imagination, informing me of the sound-maker's participation in their surroundings, and by circumstance, flavoring my participation in the fabric of their relationships.

You could say that these sounds "represent" something about my Auntie, or my neighbor. Certainly my memories of their sounds represent my experience of them—elements of which could be recreated to invoke the memories. Someone could mimic my Auntie or my neighbor and bring these memories to life, making sounds that serve as "memory pegs" to an enriched experience that the sound-mimic and I have in common. It is at this point that sound communication takes on an additional dimension; no longer an experience of one individual to the sound of another, or the experience of "some individuals" to the sounds they hear. Once there is an interlocutor, the sound experience gains perspective; the sound bears a transportable significance; it can be moved in space and time away from the original stimulus. The sound is no longer just a product of a relationship; rather it bears specific "meaning."

Humans aren't the only animals who have this ability to imbue sound with specific meaning. Predator-specific alarm calls of various animals are a good example of this. Vervet monkeys have calls that distinguish between terrestrial or airborne predators, so when a specific alarm call is heard, the troupe either heads for the trees, or jumps under them, depending on the threat.[11] Prairie dogs produce even more complicated predator identification sounds, ones that can signify size, proximity, speed of travel, and even color.[12] Researcher Con Slobodchikoff found that the prairie dogs he was studying had created a call that identified a colleague by characteristics which they recalled later, after not seeing the man for 2 months—in essence, giving the man a name.[13]

The prairie dogs (or the humans) are not unique in this "naming" skill; they are just good at it. The prairie dog's underground environment may have predisposed them to communicating in this way. In their subterranean realm, vision is low on the scale of perceptual importance. In a dark, visually limited world, these creatures have a greater reliance upon their perception of sound and vibration. These perceptions occur over time, so they stimulate memory and comparison.

From deep in their warrens it is important to know if the sounds of footfalls from above are approaching or receding; if the sound source is heavy, fast, light, or slow. These sound cues encourage the integration of time and memory—pattern recognition—conveying a dynamic sense of the surrounding conditions. This form of pattern recognition is what writer/paleontologist Loren Eiseley termed "time binding" and constitutes a possible development path to linguistic perception.[14] Time binding allows for events, memories and situations to be represented in terms of sounds—inviting acoustic mimicry and the creation of "representative" sound communication.

In the simplest of terms this involves the perception and expression of size comparisons: "anything that makes a sound larger than me is dangerous, anything smaller is not (or is a potential meal)." These size distinctions can be signified by the urgency of an alarm cry. Vocalizing animals above ground can respond to immediate threats by broadcasting their alarm cries into the open air for all to hear. But for social subterranean creatures like the prairie dogs, communication channels are limited by the setting. In their long winding tunnels there is a need to convey spatial and temporal maps to others of kin, like "don't go out the south side hole if you value your life!" Descriptions of geographic features, sources of food, and sources of danger expressed in symbolic auditory form are quite handy for them. A vocabulary of descriptive sounds allows them to carry unique maps in their minds that they can share with each other over time.

Of course while the naming behavior of the prairie dogs implies "language," it is actually just a response based communication; their calls are alarms that invoke behaviors in other community members. The integration of time is what makes it language-like, but there is no evidence yet that these guys get together after a hard day on the plains and discuss their experiences among themselves.[15] Their calls are a more akin to the shrieks of the Scrub Jays in my area when they imitate the hunting cries of the Red Tailed hawks. These imitation calls are in the vocal arsenal of the jays, exploiting a relationship that the local small animals have with the hawks. The jays seem to do it to get a rise out of all prey animals within earshot. It is not language, rather it is a sound that stimulates behaviors in the prey animals (and really gets on the nerves of the local cats). These imitations are transportable sound cells that predictably do something based on a common experience. In this case the experience is an understanding of a common threat—predatory birds from above.

Understanding the alarm calls of other animals is a variation of interspecies communication. When animals dwell in a common habitat, they become aware of their neighbor's sound behaviors. As a result, the scolding jay or the "shriek/ka-thunks" of the gray squirrel in response to a threat probably accounts for the survival of many other small forest animals. Even for aliens (like me) the alarms are not hard to decipher. As indicated in the first chapter, alarm sounds are by nature alarming. It's all of the other communication sounds that conspecifics use with each other which blur the distinction of whether it is language or not.

Group animals will share experience and perspectives that they codify into sounds. These sounds invoke the relationships and behaviors that are particular and specific to their own group. In many cases community animal sounds are not "genetically programmed" vocalizations; rather they are individual responses to their environment and community. The prairie dogs mentioned above were Gunnison's Prairie Dogs (*Cynomys gunnisoni*) in Arizona, but the naming behavior is found in other species of prairie dogs as well. They all sound like prairie dogs, but as a consequence of their particular regional experience they wouldn't speak the same dialects.

The particular vocalizations of any animal species are set in the mechanics of their body, so all Ravens (who are also community animals) will all have raven-like calls, but the "meanings" of the calls vary between groups and even between individuals within a group. Thus the Ravens in France sound much like the Ravens in

New York, but their "caws" mean totally different things.[16] Regional bird dialects can be so distinct in some cases that an individual animal's origin, upbringing and pedigree of can be determined by its regional accent. The late and fabulous ornithologist Luis Baptista did seminal work on the regional accents of white-crowned Sparrows. He became so versed in their dialects that he could distinguish the pedigree of birds whose parents were from regions displaced by only a few miles.[17]

Birds learn their songs like humans learn language; by imitating the sounds of their parents and older kin.[18] Through their learning period the young malleable minds are subject to the influence of the environment and the experience of the "teachers." This easily lends to distinct variations in pronunciation, and even meaning. In a sense, these regional accents express something about upbringing, environment, and community influence, so accents and dialects are expressions of place and relationship. Dialects can be so distinct that animals of entirely different species can distinguish the differences. It is by way of this that seals in the Straits of Juan de Fuca can distinguish difference between "safe" and "threatening" pods of Orcas. In this case, Orcas from different groups favor different foods, so when the pods that favor salmon come cruising through, the seals don't pay much heed; but when the Orcas with a taste for seals come around, the seals high-tail it out of the area.[19]

There are a number of conjectures as to why certain animals have distinguishable dialects or accents (and some don't). It could serve as a genetic marker, helping animals keep their gene pool from stagnating.[20] It may also give territorial animals cues on the sanctity of their kinship boundaries.[21] There are a number of behavioral explanations, but it could be just the way it is by happenstance, e.g., I live in California, so I speak like a Californian.

In the early 1970s I hitch-hiked around the United States a bit—crossing the continent four or five times this way. On my first sweep below the Mason-Dixon Line I was awestruck by the little kids—only 3 years old in some cases, who spoke with a deep southern drawl that was particular to the south at that time. I had thought that the southern accent was a variation on the English language that had to be learned after mastering the "non-accented" English that I spoke. I was impressed that the kids had "aced" their dialect at such an early age.

Much of these regional dialects are fading as contemporary broadcast media blurs (or obliterates) regional differences. You can still get a young Texan up on a drawl over a beer, but increasingly we all sound more like the TV. I personally hope that American regional dialects don't all dry up and disappear, because I enjoy being able to hear that folks are from somewhere else by the way they speak. But I fear that the white-wash dialect of corporate broadcasting will end up sowing the seeds of our communication using the linguistic equivalent of GMO seeds.

Perceptual Platforms

The preceding section broadens the meaning of communication and highlights it in terms of perception and response to environment. All animals—including people, use their perceptual skills to sense and experience their environment. They adapt to it in ways that are effective for their livelihood, and express their experience in ways that support or clarify their perceptions. Thus their songs, emotional expressions, reactions, memory, and onomatopoeia all enrich their sound communication. The expressions of alarm, glee, passion, pain, or delight produce sounds that we commonly understand—because these emotions bond sentient beings together in a common vocabulary of fear, joy, anger, grief, and affection.

But each sentient being also has an array of individually unique priorities which influence their vocabulary in ways that are not so easily deciphered—perceptual priorities that others are not even equipped to grasp. I am not just referring to our inability as humans to grasp the priorities of giraffes; rather I am referring to the simple fact that you and I have unique perspectives that frame our own perceptions. Specific groups of people, and even individuals within a kinship groups may have perceptual priorities so disparate as to completely flummox clear communication.

By way of example; most North Americans tend to orient our perceptions within our heads, probably because many of our dominant sense organs reside there—including our sense of balance and motion. But there are other people who meet the external world from elsewhere. The Q'ero of Peru "eat" the world from their "cozco" or bellies (from which comes the name of the city Cuzco).[22] The Diné think from the tips of their noses.[23] The Dagara of Burkina Faso have a "shadow perception"—somewhat equivalent to the manifestations of material form; a "hereness" that we can all agree with. But they also have a perception of those things that lie behind the shadow—what European minds might call "ecstatic perception."[24] For the Dagara, the world behind the shadow is the "real world," like the "dreaming" of the Aboriginal Australians—not the illusory world that we spend our waking lives in.

For the five (or six) senses that Westerners perceive the world with, the Q'ero have nine. I am not even equipped to imagine how those perceptions inform their communication. I might be able to guess at the sources of stimulation, but with the certainty of reading an instruction manual, the Penan of Borneo listen to the forest sounds as a "language" that sets the flowers to bloom, warns them away from a hunt, or clues them in on where to settle down.[25]

When we explore the perceptual framing of various indigenous peoples, we find that they are encountering the world in ways that reside outside of the European sense of space and time. Consequent time is a European perceptual priority, so we conjugate our words of action in terms of the past, present, and future (modified by levels of probability). On the American continent the Diné language (of the Navajo), is not conjugated in terms of time or the vagaries of the subjunctive tense, rather the words are all set in the sacred winds that carry all experience.[26] The Winds speak into the ears of the people giving them instructions and the words to speak. This breath of wind itself is Spirit, inspired in all life, exhaled by the speaker on its path through the legacy of being. Speaking here is not a recounting of experience or a strategic framing of intention, it is rather a continuation of experience itself; the participant becomes part of the inspiring wind, propelling the experience along as the speaker co-creates it—inferring the importance of keeping the breath and the sound of experience in play to assure its continued existence. What do the Diné hear when the winds are instructing them?

I was speaking with Tohono O'odham[27] "road man" Rupert Encinas not long ago about his language. His language is very easy on the ear, with many glottal sounds—sounds that emanate from the throat. These include various stops, clicks, "clerks" and "clocks." It sounds as if the words don't actually come out of his mouth, but rather rumble around in his shoulders, chest, throat, and sternum. He tells me that this language is closer to the heart; that even as he speaks—as his message becomes more prayerful—the sensation of his speaking sinks down deep into his body. This sound and sensation is far different than our English language, which is worked well by the tongue, the lips, and the teeth. His wife Jodi, a native English speaker, tells me that the language is very old, and that it can express very old things that cannot be expressed in English. By comparison, when we speak about very old things in English, we need to include time-reference phrases such as "in the beginning" or "there was a time very, very long ago." In Tohono O'odham, the language itself expresses the antiquity from which it arises; it is known that the expressions are formed out of the sensations and experience of the Earth.

In a similar manner the Hopi speak with sounds that are formed by relationship rather than extracted from a dictionary of set terms. Their speech is not so much a narration of consequence and expectation; rather it is an affirmation of participation in their surroundings. Like many other non-written languages, they speak more from sound-words with transitive meanings that are directly connected to what they express.[28]

What is a word with "transitive meaning"? My Kicakpoo/Sac-and-Fox Uncle Fred Wahpepah tells me about the naming of his grandson. At a naming ceremony, the Elders took this infant in their hands and passed him around. His name "Wahpepah" was available, because the Great-Grandfather who had used the name had passed on to the Spirit World. The earlier meaning of the name was "Leader of Eagles," but this did not fit this child's spirit. It was determined that his name meant "The Enticer—one who enticed the enemy into ambush." The same name, but a different message. It is from this perspective that simple names and words can "mean" complicated ideas. Fred's mother had the name "Puh'Kaa" which meant "The she-bear, just disappearing behind the curve as she walks down the trail." A name like this binds the communicators into an imaginal landscape. It acknowledges participation and assumes relationships. Speaking from this perspective also animates communication; knowing that sounds are alive and in play—serving more as mediators of common experience rather than merely representing "things."

The reciting of these sound-words stimulates the communicator's imagination in dimensions other than "associative meaning" or "representation." David Abram, in his book "*Spell of the Sensuous*" relates an account of an Apache cowboy fence-stringer who would quietly recite long series of place names to himself while working—going on for 10 minutes at a stretch. When asked what he was up to, the stringer replied that he often "…talked names" to himself. "I like to," he told the anthropologist. "I ride that way in my mind."[29]

The fence stringer's imaginal world may be similar to the world I inhabit when I listen to Ravel's "G Major Piano Concerto" or Miles Davis' "Bitches Brew"—which both seem like enchanted landscapes for me. I wish I could speak and understand a language that could transport me to these places through individual words alone, but my perceptions are not set up that way. I am dependant on constructing these landscapes with complete sentences and paragraphs (or with music, of course).

What this all speaks to is that as sentient beings, we all have an autonomous "take" on our surroundings from our own unique perspective. Each individual—each organism—has autonomous perception and response, or "perceptual platform," built over time from all of the social, biological, cognitive, and environmental inputs that the organism experiences—tailored to their own priorities. Looking at an organism in terms of their perceptual platform (as opposed to how their behaviors reconcile to a set of biological assumptions or "design criteria") opens up the possibility that each organism dwells in their own perfection, and not somewhere along the developmental path to becoming a "highly evolved modern human being."

With this perspective—along with our fertile imagination, we can explore the communication realms of other people and other animals to broaden our own perceptual platforms.[30] Because we participate in mutually inhabited acoustical spaces, sound serves as a handy bridge between us and others in our environment. From this broader meaning, "interspecies communication" amounts to an organism's ability to adapt to the communication cues of other creatures: Thus the rabbit knows what to do when it hears the alarm calls of the gray squirrel, and humans can understand the emotional content in the vocalizations of their dogs.

There will always be limitations to developing a mutual language with others, determined by differences in our social and environmental perspectives; or in the case of other animals, with our unique biological faculties. Dogs don't have human lips, humans are unable to hear the vocalizations of bats, and spiders speak with vibrations through their webs—a communication channel that we are not adapted to. Even when we do have common communication channels, clear communication is limited by our unique experience and priorities. So while I can clearly hear the vocalizations of a Tohono O'odham road man, and I could possible mimic the sounds I hear, my ability to understand what he is really saying is constrained by more than just my not understanding his vocabulary, it is also limited my perceiving the world differently than he does. His perceptual platform is built from different experiences, different priorities, even different concepts of time and space than my own.

Communicating between perceptual platforms can be a challenge even when people speak the same language. For example—again from my Uncle Fred, who was once courting a girl but was concerned about her honesty because she wouldn't look him in the eyes when he spoke to her. He asked his cousin for advice on this. His cousin told him that she had probably been raised by her grandmother—and that the Elders like her grandmother wouldn't face you when they were conversing, rather they would turn their ear toward you so they could hear you better.

Fred's misunderstanding was cleared up and he eventually married his beloved. But this misunderstanding was due to a difference in convention; the grandmother who raised Fred's wife gave priority to an auditory communication channel, Fred's courtship priorities were more visual. He was able to reconcile the differences once he was informed about them because they did have other similarities in their perceptual platforms. But even where the platforms have similarities, communication can be thwarted by differing perspectives. I know that I inhabit a different perceptual platform than a Native from the jungles of Malaysia, and I would account for this while attempting to communicate with him. But I may have a harder time reconciling the perceptual differences between me and another middle-class white American, particularly if I assume that we do have similar perceptual platforms.

Framing communication in terms of perceptual platforms establishes autonomy for each communicator. It also highlights the fact that sound communication really depends on a process of participation, or "listening." Studies and evaluations of speech intelligibility typically assess a listener's ability to discriminate meaningful words in a background of perceptual interference, such as reverberation, masking noise or gibberish.[31] But there are many other factors that influence a listener's ability to understand, such as context, familiarity with the speaker, word and sentence complexity, speed of delivery and so forth. Given the importance of understanding speech to human society, the field of speech intelligibility is rich with studies, examples, and methodology, so even a cursory glance would be well beyond the scope of this book. But if you consider that a simple sentence such as "Please slide the book my way" involves action, command, social hierarchy and comportment, context, discrimination, objects, dependant clauses, and sound, there is a lot of information loaded here, and lots of room for interpretation.

Humans can grasp much of what comes into their minds by way of their ears, but often there is a lot of filling in. Amazingly, sentences constructed out of as much as 80 % unintelligible nonsense can be sorted out into an accurate sound communication. Filling in that 80 % requires a lot of assumptions on the part of a listener, and in this there is a lot of latitude for interpretive error. 80 % is the extreme case, but to some degree all meaningful sentences are extracted from a stream of sounds composed of proto-words melted and mumbled into each other. A familiar example is the "Jose can you see by the donzerly light" ditty—which is one representation of how we hear a familiar sentence.[32] Fortunately the mind has the tools for sorting out phonemes into words and sentences. Experientially these tools are based on learning, powers of association, deductive reasoning, and a library of memories. There is also mounting evidence that there are neurological tools for sorting out phonemes as well—brain processing specific to sorting streams of sounds out into strings of words.[33] Effective verbal communication between two people assumes that their listening tools intersect at some points, and that there are common perspectives and even common processes of cognition between the speakers. It assumes that both speakers can recognize each other's perceptual platform. This is a pretty big assumption, and while it does work most of the time, it is amazing what can happen when it doesn't.

Most of our verbal communications do end up stumbling along at a reasonable translation rate, but to get an idea of how well we really understand each other, there is an exercise employed by professional mediators and therapists that involves "listening feedback" when two people are attempting to convey their feelings or intentions about each other. The technique involves giving one person the role of the speaker and the other the role of the listener. After the speaker makes a statement, the listener tells the speaker exactly what they heard the speaker say. Almost invariably—at least out the gate—the speaker has to reiterate or restate their original sentence to clarify what they originally attempted to convey.

This exercise really isolates the "language" part of the communication from the sound part and demonstrates that complete communication involves more than just words. Tone of voice, speed of delivery, body language, and cadence convey much of the communication. These are all subject to interpretation, based on the expectations of the listener, and where much of the success of communication resides—regardless of the actual words used.

Clarity in communication assumes that the participants have a common lexicon—some common experience and some common assumptions about their community or society. It also assumes that the perceptual platforms are stationary, which is not necessarily the case. Emotional dispositions are constantly shifting the perspective; depression, joy, anger, satisfaction, and confidence will all affect perspectives, and thus perception. When we get to know someone, we learn the range of their perceptual platforms, and they become more predictable to us.

But there is also an unpredictable element of perception that rarely gets folded into the mix; this is when the communicators have distinct neurological or cognitive differences that can't be so easily qualified. Perhaps the most common cases involve what is known as "dyslexia." Until recently dyslexia was considered a visual disorder that impaired reading ability. The common understanding involved the belief that the letters of written words were somehow scrambled between seeing them and perceiving them. It was believed that somewhere along this neurological path the dyslexic flipped or inverted letters. But recent studies reveal that dyslexia is more of a perceptual problem than a neurological one. One model portrays it as a "phonological processing disorder"—the inability for dyslexics to sort out phonemes—the individual sound elements of language.[34] Dyslexics don't necessarily have any problem talking or hearing, it's just that they process sound differently than "non-dyslexics." For dyslexics, word sounds have autonomy distinct from their "meaning." For example; while they will be able to clearly describe in detail the difference between a "tornado" and a "volcano," they may not immediately be able to summon the proper word for each when needed. In order to accommodate for this perceptual variation, they may come to comprehend tornados and volcanoes, each in minute and exacting detail. Because they can't easily reduce the "tornado phenomena" into a sound-symbol, a dyslexic may have the advantage of really understanding it. This might partially clarify the exceptional cognitive skills of W.B. Yeats, Albert Einstein, and George Patton, who were all dyslexic. This is not to imply that all dyslexics are brilliant, but given that dyslexia affects 20 % of us, it may be more appropriate to consider it a perceptual paradigm, rather than a handicap.[35]

But in extreme cases dyslexia can be a real problem. If people have a difficult time grasping meaning out of the sounds of words, listening, reading, and communicating become a challenge for them.[36] A father of a close friend was so dyslexic that verbal communication was incredibly difficult for him—so difficult that he eventually gave up listening or trying to communicate altogether. Essentially he surrendered and became "dyslexicly deaf," isolating himself in his own world of nonverbal thoughts. This was exceedingly frustrating for my friend, as she knew her father could hear, he just couldn't hear *her*. This came to a dramatic apex when they were hiking together in the Sierras. She was quite a bit ahead of him on the trail, and when she looked back, she saw a huge boulder rolling down toward him from above. She started screaming to get his attention, but he was so locked in his reverie he didn't hear her alarm. He did finally hear the boulder tumbling down upon him, but his life was just moments in the balance.

I haven't checked with the pros, but I am probably "dyslexic" to some degree. I attribute the way I hear sound to my being a musician. One indicator of this is that when I hear songs I really need to concentrate to hear the lyrics—even with straightforward tunes like the music of the Beatles. My perceptual platform tends to hear the music of communication, not the specific words. When Bob Dylan was at the height of his popularity, I couldn't understand what the big fuss was all about. His singing voice was not exceptional, and his harmonica playing wasn't particularly virtuosic. Eventually at the insistence of a respected friend I sat down and focused on the words that I had been missing, finally understanding Dylan's genius.

What this all speaks to is the fact that we are all individually adapted (and gifted) with specific tools of perception. When it comes to reaching out to others to convey our perceptions to them, the fundamental limitation is that we are unable to really understand their perspective because we are not them. We are none the less held together by inference, mutual agreements, and common responses to situations—which we confirm by our physical presence and our continued desire to communicate.

The bulk of the preceding discussion involves communicating with other humans, where the cognitive disparities both thwart and enrich our shared experience. When we evaluate the perceptions and communication of other species, the "perceptual platform" idea can help us if we frame the inquiry in the context of the other animals' priorities, rather than from a set of criteria that we humans find useful. It may help us understand the many ways animals communicate without specific words. Part of our challenge is that humans are poorly adapted to perceive the spatial, olfactory, or electrical communications of other animals. But sound—because it is such a deliberate mediation of common space, provides us with many clues to how animals convey their priorities to each other.

Acoustic Community, Acoustical Niches

Behind my house there is a little patch of land that runs a bit wild. I scratch away at it occasionally to plant some tomatoes in the spring, but I also let the grasses and fennel grow unmolested. This last spring I was down on my knees clearing away the weeds and noticed a delicate jingling sound around me, almost as if the soil was shimmering. Exploring this a bit closer I found that the sounds were being produced by tiny little ground crickets about the size of melon seeds.[37] Their song was so delicate that by standing up I nearly came out of range—the surrounding soundscape at standing-height being louder, it masked the soundscape down in the tangle of weeds at ground level. This gave me a bit of a thrill; I felt as if I had found a whole new world parallel to my usual one, but in a different, Lilliputian dimension.

I thought about these whispering animals; the tiny crickets, the equally small wasps and gnats, and ants who sing a delicate song by clattering their jaws.[38] In this little geography, the animals are only concerned with a range of a few meters—at a sound level so quiet that their music is mostly missed. I wondered if these animals were able to hear the larger soundscape around them; the sound of jets overhead or my telephone ringing in the house. Were these much larger but distant sounds part of their miniscule soundscape? Or were they unheard by them, like the music of the spheres is for us?

Canadian composer and author R. Murray Schafer introduced some important terms into the discussion of bio-acoustics—words having to do with the subjective experience of sound. "Soundscape" from his coinage, has become indispensable in describing our auditory surroundings. An equally important idea is the "acoustic community" which describes the inhabitants within a geographic boundary defined by "soundmarks" heard in their surrounding horizon, though it also refers to the common acoustical experience of people with a shared perspective of the sounds that they hear in common.[39]

Schafer uses these terms in the context of human experience; how the lowing of cattle and the clang of their bells speaks to the farming community in a way that the drone of the foghorn and the clang of the bell-buoy speak to those living by the sea. The common experiences of regional sounds have a binding effect on those living in their respective soundscapes. These sounds speak the patterns of their lives, which in turn influence their sense of belonging and continuity—a communication with their collective environment unique to those who live in it.

It is in a similar manner that the tiny ground crickets, ants, gnats, and wasps also inhabit an "acoustic community" of sorts; their sounds mediated by habitat and affected by each other's sounds. These animals handle their soundscapes differently than humans, given that they have different priorities. In an environment that may be visually constrained, such as in the tangle of weeds in a forest or jungle, or open to wide vistas, such as out on a broad savannah, resident animals listen to each other for survival cues. If they vocalize, their vocalizations are adapted to their environment, using sound channels that are uncluttered and available, using frequencies and textures that most effectively convey their sounds to the intended receivers. So song birds who dwell at differing levels in the forest will sing at pitches and textures adapted to penetrate the sound absorbing characteristics in their particular habitat,[40] and deep diving whales use ultra-low frequency sounds to reach out across the ocean depths.

Nature sound recordist Bernie Krause noticed these adaptations across all vocalizing animal species and expressed it with his idea of "acoustical niches." He noticed that "all sound-producing organisms generate sounds that fit uniquely into their environment relative to other vocal or sound-creating organisms in that territory." He also noticed that a healthy acoustical environment is like a well balanced orchestra, with all frequency bands proportionally occupied by the vocalizations of the animals in it.[41]

If we delve into this idea a bit further, we find that acoustical niches are not just dependant on frequency and amplitude or confined to environmental characteristics.

As indicted in the previous chapter, animal sense organs are adapted to many other dimensions of acoustical energy and the way it plays into their environment. Sense of size or scale, time of day or night, time of year or season, location (forest, plains, jungles, coastal shelf or deep oceans), rhythm cycles, time/frequency domains, and spatial domains all figure in the orchestra of natural sounds.

Human sound perception intersects some of these realms with our acute pitch discrimination and our sense of time and rhythm, but we are less sensitive to other dimensions. For example; we only have a rudimentary ability to discriminate phase and fine-scale time-domain cues (relative to cricket and cicada perception) and we are unable to differentiate the particle motion and pressure gradients of sound important to some fishes—or even hear the complex ultrasonic signals of bats and dolphins. And at the low end of the sound spectrum, we are just beginning to find a surprising density bio-acoustic activity which we only call "infrasonic" because it is below our ability to pitch-discriminate.

There are many bio-acoustic niches awaiting our exploration—some more apparent than others. This is particularly the case in the sensual realms where the experience of habitat and the consequent adaptations are too slow or too fast for our sense of time, or the tones are too low or too high for our perception of frequency, and where there are important biological patterns and pitches that we are not equipped to perceive.

Infrasonic Communication and Acoustical Cohesion

Across the spectrum of bio-acoustic activity, the "infrasonic" frequency band is perhaps the least examined niche. This is probably due to that fact that in addition to not being able to hear the frequencies of these sounds, we are also unable to really hear them in our own time domain; they need to be recorded and "sped up" in playback to bring them into our perceptual grasp. We can occasionally sense these low frequencies, but we tend to feel them in our bodies rather than "hear" them through our ears. The sensation of infrasonic energy gives us cues about large bodies moving—like the movements of huge predators or earthquakes, so we "feel" it as an unaccounted-for anxiety—or a "caving in" sensation in our chests.

We don't use this acoustical niche much, but many other animals do. Tigers use infrasonics as a hunting strategy, paralyzing their prey in terror before striking.[42] Alligators use infrasonic sound in courtship and territorial signaling with bellows so loud that it sets the water surface around them "dancing."[43] They may also use it like the tigers, to paralyze their prey, but to my knowledge this hasn't yet been explored.

Giraffes and their horse-like cousins, the okapi,[44] also vocalize in this range. Until recently these animals were considered mute, communicating exclusively through visual channels, or perhaps "psychically" as inferred in the popular idea of "horse whispering."[45] Now we are finding that horses also communicate through infrasonic sound.[46]

We have already discussed the use of infrasonic vocalizations by whales, who communicate through ocean sound channels with long wavelength sounds that adhere to the curvature of the earth. If we consider various animal habitats, the usefulness of low frequency sound can clue us in to where else we might find other infrasonic, or at least low frequency communication channels. Intuitively we might expect these long wavelengths to be produced by large animals, and they are: Whales, elephants, horses, rhinos, hippos, and alligators all use infrasonics, but there are many small animals that make use of low frequency sound as well. Some produce it externally, like kangaroo rats[47] or woodpeckers,[48] who both have repertoires of percussive sounds produced by striking to convey specific meanings. Other small animals produce their sounds internally like the gray squirrel's "shriek/ka-thunk." I haven't found the physiology of the squirrel's projecting mechanism in the literature, but in most North American forests their ubiquitous sound can be heard through the foliage and around tree trunks at distances of a 100 feet or more.

Many birds produce low frequency vocalizations, which we are familiar with from the glottal purrs and guttural noises of pigeons. In the parks and plazas the pigeon's infrasonic sounds project out into the open and are fairly innocuous if they are heard at all, but if you have ever suffered pigeons roosting in your attic, their infrasonic energy in the confines of your home sounds like they're tearing the house down. The inflating chests of the male pigeons are probably the source of this sound by way of some modified bladder. If this is the case, pigeons vocalize in a manner akin to some other birds, like the male sage grouse or the frigate birds (with their visually stunning gullet pouches). These birds rhythmically inflate and deflate these air bladders, producing a complex array of low-frequency sounds.[49]

A variation on this bladder theme belongs to the marabou stork. The marabou is the largest of the stork family and lives in parts of Africa, with a smaller cousin in southern Asia. With the necrophilic dining habits of a vulture and its own dour looks, it is not a pretty bird—suggested by their collective pronoun, a "muttering of marabou." This appellation may have as much to do with their appearance as with the infrasonic sounds they produce. We can hear the stork's gurgling or "muttering" produced by way of a large pouch dangling below its long conical bill. This pouch looks like a scrotum and contains a system of air sacs connected to their left nostril (how precious!).

Infrasonic vocalization is not limited to birds and large mammals. Many frogs also have the equipment to generate infrasonics with their inflatable throats. In fact, any animal with an inflatable pouch or bladder is suspect, from bladder cicadas to anole lizards. Many fish use their swim bladder as a sound generator,[50] and quite a number of primates do as well, including the orangutans, siamangs, chimpanzees, howler monkeys and apes.[51]

Because low frequency sound communication transmits sound over relatively long distances, it allows individual animals the ability to reach out beyond their line of sight, over terrain, and through foliage. This advantage is conferred to the community in allowing the whole community to keep tabs on their kin over a large distribution range, a feature particularly useful for the huge foraging animals, such as elephants or baleen whales. These animals eat vast quantities of food. If they always needed to be close together for communication purposes, they would likely wreck havoc on the environment with high concentrations of local food depletion.

With their long wavelength, long distance vocalizations, they can spread out for feeding, and come together seasonally for breeding, calving, or other community business.

Long wavelengths serve another purpose in as much as the sound field doesn't just impinge on the ears, the acoustical energy literally envelopes the body. This visceral aspect of low frequency sound accounts for how the low frequency component of music sets the body to dancing. The energy is felt as much as it is heard, bringing the organic rhythms of heartbeat, footsteps, and brainwaves into sympathetic motion. Cognitive scientist William Benzon has suggested that this mechanism serves a biological function of encouraging dancing, which in turn promotes community cohesion.[52] And while dancing is considered a voluntary activity for humans, it may well be a biological imperative for other animals. The synchronized swimming of schooling fish was mentioned in a previous chapter—how the school swims almost as a single body; undulating to the low frequency pressure-gradients, co-generated by all of the fishes in the school and tempered by the surge and flow of the surrounding currents. The fishes' community cohesion into a school serves as a defense mechanism against predation. In a school formation, the low frequency pressure wave from an approaching predator is transmitted to all fish in the school, allowing the whole school to make coordinated evasive maneuvers—hopefully skunking the predator. Individuals in the school less fit to follow the school motion (through disease, age, or adaptation) will be winnowed out sooner, pruning the less fit and encouraging the evolution of low frequency sensitivity and school cohesion.

Flocking birds also show an adaptation to acoustical cohesion through low frequency sound perception. This is particularly evident with the shorebirds—the plovers, sandpipers and killdeers, whose tight flocks dart and flash around the invisible currents of the shore breezes. The synchronization of their motion—again as an apparent single body, serves the whole flock with an aerodynamic advantage not available to a single bird.

The flapping wings of birds are a natural infrasonic generator. Exploring it as a communication channel is likely to yield significant information about nonvocal sound communication in birds. Flapping wings are so much a part of bird behavior it would be surprising if nature didn't design it in to their auditory sense of space. There is naturally going to be some noncommunicative flapping during flight—just getting from here to there, but I have often noticed a more forced fluttering or flapping from perching birds as they move around each other or around other animals.[53] While it is not strongly apparent in most birds, wing-generated sound communication is really evident in the breeding displays of hummingbirds. During breeding season, the male hummingbird will execute a series of extreme power dives over a prospective mate. Soaring up to 50′ above the female, they dive down with a forced fluttering that produces a howl toward the lower area of their dive. The volume of this howl is impressive to me as a bystander; I can only imagine how it impresses the ladies—being the acoustical focus of the maneuvers.

The hummingbird's howl is not infrasonic to human hearing, but it is "low frequency" to hummingbirds, given the size and vocalization range of these little birds. The wing-generated sounds of larger birds are considerably lower in frequency.

Geese, for example—being among the largest of flocking birds—have a wing-beat frequency of around 2–3 Hz. When I lived in New York's Hudson Valley I would see the geese coming in through the late fall—just in front of the deep winter storms from the North. The geese travel in flocks of hundreds of birds, cutting their deep V's across the sky. Often you would know they were approaching before you saw them—you would sort of "feel" them. Their approach was palpable. The density of the atmosphere would shift and the next thing you would witness was this apparition pressing across the gray autumn sky.

"Seeing" with Sound: Ultrasonics and Echolocation

It was Leonardo da Vinci who in 1490 first suggested that remote objects could be detected with sound.[54] His scheme involved placing a pipe underwater and listening through the tube end for distant ships. 300 years later Lazzaro Spallanzani and Louis Jurine determined by experiments that bats used sound to see, though they didn't quite know how.[55] It wasn't until 1906 that H. Hartridge proposed that bats used "short wavelength sound:"

> …. The sense concerned must be able to perceive objects when they are still a considerable distance away. This conclusion would appear, by the exclusion of vision and touch, to have brought us to the third conclusion, namely, that bats possess some sixth sense not found in the case of man. I am not however prepared to accept that conclusion until another explanation has been tested and proved to be at fault, namely, that bats depend on their hearing for the directional control of their flight at night.

…and after a few paragraphs about the physical qualities of high frequency sound:

> I suggest then that bats during flight emit a short wave-length note and that this sound is reflected from objects in the vicinity.[56]

It still was another 30 years before Donald Griffin confirmed that bats used high frequency sound to perceive their surroundings.[57] Griffin was familiar with both Spallanzani and Hartridge's work, though unlike his theoretical predecessors, he had access to the equipment to confirm the presence of high frequency vocalizations.[58] Even at this point the idea seemed radical and required more than published papers to convince people. In Griffin's later words: "Radar and sonar were still highly classified developments in military technology, and the notion that bats might do anything even remotely analogous to the latest triumphs of electronic engineering struck most people as not only implausible but emotionally repugnant."[59]

Now, of course we are quite familiar with the idea of "ultrasound" and high frequency sonar and use the physical characteristics of short wavelength sound to derive fine sonar detail and penetrate tissues to scan the inner workings of our bodies. We use "ultrasound" to "see," just as bats, dolphins, and some birds do (e.g., the

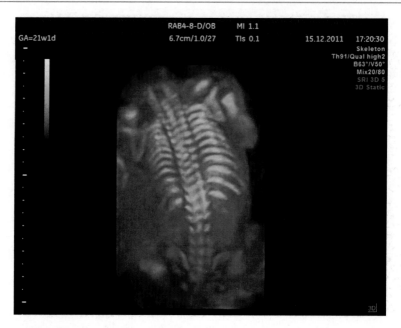

Fig. 5.1 3D sonogram-spine (Photo Wolfgang Moroder)

cave shrike and the oil bird). We think of ultrasound in visual terms because it lends itself to conversion into visual displays—from fish-finders to medical imaging.

Our technologies convert the high frequency reflected signals into screen images with colors that represent location and density. Though it is hard to imagine that echolocating animals "see" a color soundfield, it is even more difficult for me to imagine what they do perceive as they precisely navigate their environs of sonic density gradients and acoustical opacities. In many ways it might be the inverse of what we are accustomed to when we see hard reflective objects with their distinct boundaries and outlines. In our visual realm we see walls as clear, solid objects; edges are clean with precise boundaries. But in the realm of high frequency bio-sonar hard objects with sharp physical boundaries will reflect and scatter ultrasound into a fuzzy array of multiple specular reflections, yielding soft, diffuse "images" of their complex reflective sound-fields. Meanwhile soft sound-absorbing objects which don't reflect sound well would reveal distinct and clear reflections of their forms.

So when a dolphin ensonifies a human swimmer with their ultrasonic bio-sonar, the acoustical reflection of human tissue and hair would melt into the surrounding water—embraced by a semi-diffuse field of bones, muscle, and teeth, and punctuated by the fuzzy echoes of buttons, zippers, and snaps. Similarly bats flying around in a cave navigating by way of bio-sonar would avoid walls not by their etched and incise acoustical presence, but by their scattered diffuse sound reflections.

Considering bio-sonar in this way helps us understand what an individual bat hears when it joins thousands of others on their nightly cave exit to forage for food. With all of the chatter in the same frequency band they need to lock into the specific reflections of their own voice to avoid collision with others in their swarm. They must have some time-domain filter to sort out the needed reflections from the gargantuan fountain of sound of which they are a part—or the sum of all reflections from their conspecifics provides a general "acoustical illumination"[60] of their soundfield which they can all navigate within.

We know that bats live in a realm of high frequency sound reflections, which some of their prey can defensively hear.[61] We also know that dolphins and porpoises work the ultrasonic realms, and in defense of their dolphin predators so do some of their fish prey.[62] One of the mysteries in ocean bio-acoustics is how fish and other marine critters navigate at night when their visual acuity is limited by darkness and their acoustical repertoire does not include active bio-sonar.

One theory proposed by Dr. John Potter suggests that the ambient environmental noise—particularly the broadband noise generated by the ubiquitous snapping shrimp informs fish and other marine animals about their surroundings [63] (see Chap. 4). His conjecture is that ambient noise generated by the shrimp provides an "acoustical daylight" that "illuminates" the surroundings allowing marine animals to hear where they are much in the way that sunlight or artificial light sources illuminate our surroundings allowing us to see where we are.

But given the existing evidence of the hearing frequencies of most fish, much of the high frequency component of the snapping shrimp's "acoustic daylight" would be outside of their hearing range. But if they can get around in light conditions that are too dark to stimulate their visual senses, do they process their spatial-sensory information through other channels? If they are sensing the acoustical energy of their surroundings, is there a perceptual transition between voluntary and autonomic responses to sound stimulus on the frequency continuum that is not being revealed by the common threshold testing procedures? These and many other bio-acoustic questions remain to be explored.

"Ultrasonic" and "infrasonic" acoustical energy are by definition outside of the human hearing range, but they categorize "frequencies" of acoustical energy which are physical properties that we can measure. "Frequency" expresses repetitions over time—how many cycles something occurs over a particular time interval. The reciprocal of "frequency" is how much time unfolds between a set of two or more repetitions. While this distinction may seem academic, it actually engages two different cognitive abilities. One is the ability to integrate a stream of repetitive events into a perceived quality called "pitch," the other is an ability to distinguish time intervals between a sequence of events—in the "time domain." Humans have the ability to do both to some degree with our ability to discriminate pitch and timbre on the one hand, and our ability to grasp and synchronize to rhythm on the other.

Human "time domain" perception of acoustical energy picks up at some long pulsed interval in time, such as the thump of a slow turning wheel or the long beat of a drum—perhaps one pulse every few seconds, below which we hear only single events.[64] As the pulses increase in frequency we hear the rhythm mount, then melt into a rhythmic flutter up to about 20 cycles per second, at which point we begin integrating the pulsed events into pitch perception—a hum or a growl which we can hear as a tone. As we sweep the tone frequencies up we hear it as pitches until it approaches 17,000–20,000 Hz, after which point we do not hear any sound at all. So it is by way of the limits of our pitch perception that the terms "ultrasonic" and "infrasonic" exist.

This integration of frequency into pitch is a perceptual feature that is facilitated by the cochlea—an organ unique to mammals. Most other animals without a cochlea may be able to discriminate pitch to some degree; but given that they don't have an organ that specifically sorts sound out into pitches it is likely that they perceive sound mostly in the time domain.

The time domain of stridulating insects

There are many insects that make noise of some sort or another. There are butterflies that hiss like snakes,[65] and beetles that chirp, hiss, or crack.[66] There are even some caterpillars that scratch out tunes on their leafy substrates.[67] But when most of us think of insect songs, we think of stridulating insects; the crickets and cicadas, katydids and grasshoppers. "Stridulating" refers to the way these animals generate sound by way of physical oscillation, with files and plectrums on their wings, bodies, or legs, or in the case of the cicadas, a set of diaphragms or "timbals" (this structure is akin to the little metal dimple-click toys that snap when pressed and snap again when released). The various mechanisms of sound production are all ingenious, and the specific ways each species use them are equally diverse. While the mechanisms are intriguing in their own right, for the purpose of this discussion I'm going to focus on how they are used for communication.

As I excavated the literature on stridulating insects one of the first things that popped up is how niche-specific these animals are. While the sound of summer crickets may seem ubiquitous, often what we are hearing are different species, each in their own specific habitat, or taking species-specific "song shifts" throughout the day and night. Certain species will only inhabit moist bogs, others only in dry leaf litter; some sing only in the early spring, others only in late July. The afternoon singers are different than the dawn singers, and even when species sing at the same time at night, their songs will be a slightly different pitch or at a slightly different tempo.[68] When I listen to the jingling and pulsing of country crickets on a summer night I can often differentiate as many as three or four different songs; one pulsing a high ring, another pulsing a lower tenuous jingle, a third with a continuous tinkling or trilling over a background of a sweet metallic purr of yet another species.

The reasons for the various sounds of stridulating individuals are typically described under the triumvirate of behaviorism: predator evasion, territorial defense, and mate attraction. But the common behavior that ties all of these stridulating insect communities together is their chorusing. Chorusing behavior is usually framed in the context of fear; that animals synchronize to avoid predator detection i.e.: "I will not risk making a sound unless I am covered by the sound of others." I'm sure that there is merit in this, though I believe that chorusing is more like the flocking of birds or the schooling of fish in that while it may be tempered by fear, it requires cooperation. From this perspective the incentive for cooperation falls under the rubric of "acoustical cohesion."

Unlike the birds and fish, chorusing insects are relatively stationary; they don't move as a group, rather they dwell in a common habitat. Any movements within their habitat are specific to individuals who wander about a relatively small area attending to their feeding and breeding activities. Their stationary community requires various strategies for community cohesion and predator evasion. Predator evasion by way of producing an ambiguous soundfield was explored in the previous chapter, but a second evasion strategy dovetails into their acoustical coherence as a community.

When a community of stridulating insects is pulsing, buzzing, or trilling away, they are advertising their presence, but they are also locking into the community body. If one insect quits singing (due to sensing a potential threat, for example), the others are alerted to the absence of the individual sound, so they respond accordingly. In some cases when one animal quits, it shuts down the whole operation; in other cases the others maintain their song, perhaps modified in some subtle way in response to the absent individual.

Most of the literature on community synchronization and chorusing frames this behavior in the context of competitive mate attraction strategies.[69] This is inspired by the fact that in the majority of cases, the singers are males. Many of the studies examine the various aspects of female attraction to a call's characteristics such as volume, frequency, "first callers," proximity, and such, so undoubtedly chorusing does play a role in mate selection. Individual territorial domination probably also influences chorusing, given that individuals are located somewhere within the body of the chorusing community but still have their own genetic priorities to tend, although I suspect that some other things are happening as well.

A few summers ago I was out in my yard taking in the heat of the late morning. The cicadas were buzzing away, letting me know that the afternoon was going to be even warmer. While being aware of the buzzing soundfield evenly distributed around the yard in and among the lower branches of the surrounding trees, but I wasn't paying particular attention to it. Suddenly, and for no apparent reason the cicada (or cicadas?) in a ginkgo tree ceased singing. This tree is set apart from the rest, so silence from that quadrant was really apparent. I didn't notice it as a silence though; rather I was jolted by a momentary sense of vertigo—almost as if someone had pulled the rug out from under my right ear.

It was immediately clear to me what had occurred; the cicadas buzzing around my yard were phase-locked to each other and I had entrained to their buzzing, stabilizing my outdoor sense of space to their soundfield. When half of that soundfield collapsed, it was almost as if half of my hearing was put into freefall. Needless to say I investigated this further. I regret not having more appropriate equipment to delve into the minutiae, but I did unravel this mystery a bit.[70]

These particular insects were buzzing at a fundamental frequency of ~14 Hz, but the harmonic content was rich and extended well up into the ultrasonic range above 20 kHz. I was able to determine that their fundamental 14 Hz was synchronized on a fine scale, and I can assume from my sensation of vertigo that their phase synchronization extends up into the harmonic range of my own localization cues—between 2 and 7 kHz. This would account for my "spatial entrainment" to their soundfield.

How high up their phase lock occurs would be a matter worth exploring, as the subtle time-domain fluctuations in their harmonics may serve as a spatial cue for them, carrying information about the shape of their collective acoustic domain. If you can imagine their perceptual platform, these creatures dwell in a limited visual realm. Their input priorities are close range; they need to sense the motion of the surrounding foliage swaying in the breezes, differentiating these motions from the motions of large nearby predators. Meanwhile, they are somewhat saturated in their self-generated and conspecific-generated soundfield. I almost sense their soundfield as a cloud-shaped geography, modulated by the external forces that impinge on it—much like the fluid body of a school of fish. Subtle changes in the perimeter of the soundfield would inform the whole community body on the nature and form of their surroundings—from incoming predators to the texture of the prevailing breezes and thermal turbulence. If you could visualize their soundfield, you

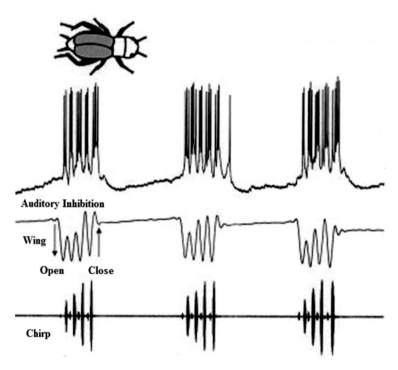

Fig. 5.2 Corollary discharge in crickets. Inhibitor neurons blank auditory sound perception in phase with chirp syllables thus the cricket does not deafen itself while cricking, rather it hears "nothing" during it's cricks[76]

might see a large amorphous cloud, undulating and pulsing in response to the motions and influence of the surrounding environment.

While this may seem overly speculative, the orthoptera as an order do have the equipment to discriminate subtle phase cues and resolve fine-scale time domain information.[71,72] They also have various methods of phase-locking to external signals.[73,74] The cicadas have timbals on each side of their bodies. If they are able to phase modulate them independently, it would account for how they generate ambiguous signals that are hard to locate. It may also present a method of modulating the fine-scale time domain of their song, phase-locking into their community soundfield.

When I had entrained to the cicada's sound field in my back yard I was probably placed in a geometrically opportune position—perhaps equidistant between two synchronized animals, so the effect of their synchronization was really pronounced.[75] From the perspective of each individual animal, they would need to synchronize their call with a distant and relatively much quieter call. This poses a little problem—but it has a tidy solution.

Stridulating insects are most sensitive to the very sounds that they make, so in calling, they run the risk of deafening themselves. In order to protect their own hearing they have a way of inhibiting or "blanking" their auditory neural activity simultaneously with their call.[76] Blanking prevents them from being deafened by their own sounds and would coincidently provide them a precise time domain cue with which to synchronize: Any sound outside of their inhibition phase would be heard, any sound simultaneous with it would not, providing them with a "negative auditory target" that would be much easier to hit than a "positive" target.

With this method of synchrony stridulating insects can dwell in a community body "neurally synchronized" to the fundamental pitch of their call frequency, while not really hearing the pulse of the community call. What they would hear are the little chirps and blips of their kin (and of course any other sounds) that fall outside of the community-synchronized blanking interval. These asynchronous chirps and blips might occur when an individual insect shifts its call in response to some external stimulus like an approaching predator, the motions of a breeze through their habitat, or the approach of a nonthreatening animal. As you or another potential threat approaches the perimeter of their community soundfield, those insects close enough to sense your presence will respond by a tentative slowing down or halting of their stridulation—alerting the rest of the community body about your presence, and in some cases causing a complete halt to the entire chorus.

What this lends to is the idea that stridulating insects inhabit an acoustical space that is like a "dynamic map" of their surroundings. This map would inform all individuals within the body about the conditions of their surroundings, giving each one of them information that they would not otherwise be able to ascertain without the input of the others. The concept of "auditory maps" provides us with another way of considering sound perception as a "perception of placement and orientation." The idea also frames another way we communicate with, and hear our own environment.

Mapping: Memory, cognition, and participation

Typically when we think of maps we picture a chart that represents some terrain; an intersection between a scale model of a territory and the icons, words, and symbols that help direct our thinking about the model. When we refer to a map for way-finding we inhabit two realms; the realm of our physical placement, and the realm of our imagination—which navigates the map's portrayal of a terrain. Maps are interactive. They are much more than a set of serial instructions to get from here to there; rather they are expressions of relationships.[77]

While many of us think of maps as visual products that we can "toss into the glovebox" or access on our smart-phones, mapping is a large part of how we navigate our surroundings. It is a cognitive process that translates experience into orientation, taking note of details and characteristics and laying a foundation for our sense of place. The details and characteristics we place on our cognitive maps are much more than visual landmarks put on a geographical grid, they are more akin to embellished dreamscapes; adorned with our attractions and fears, placed among our memories of the past and our hopes for the future. We dwell in this map, carrying it with us as an ongoing adaptation to our environment. As we become accustomed to new surroundings, we map our relationships with it on a dynamic scale of importance, giving experiences that are common and familiar a different perceptual priority than those that are unique and new.

We often consciously locate things in our sphere of influence by seeing them, but we locate ourselves in our surroundings by many other artifacts of memory: Pain and joy, odor and feel, and of course, sound. As we have previously explored, sound helps us establish the extents of our sphere of influence and gauge the boundaries of our community. It helps us identify what is downwind from our sense of smell and just beyond the reach of our vision. We readily hear which sounds are new, and easily habituate to the regular soundmarks of our setting. These reactions and responses constitute our lexicon of familiarity and belonging to our surroundings.

Our maps are necessarily complex, bearing the generalities of human experience and culture as well as the artifacts of our personal perspectives. While our maps are obviously more complex, they parallel the dynamic auditory maps of the stridulating insects—which also summarize their sphere of influence and community boundaries as a product of their collective perception. While I am reticent to call the mediating inputs to the insect's map a "collective experience," in its simplest form it is just that; a group response to their experience of their surroundings. There is a biological imperative for these insects just to buzz, jingle, or crick away, but there is also enough cognition and memory in their behavior for them to adapt to changes in their setting. If this were not the case, my little cricket friend mentioned earlier (Chap. 4) would not have been able to recognize me as a non-threat. If the crickets were not able to qualify threats on some graduated scale, their community would all cease stridulating anytime their static setting changed beyond a certain threshold—for fear of some unknown, unrecognized, or unmemorized threat. Their collective memory becomes a shared cognitive map within which they all comfortably dwell.

It so happens that these cognitive maps are what most sentient beings create and dwell in. Mapping is how we avoid having to evaluate every bit of stimulus as "new information" to reckon with. These maps are products of sensory inputs that tend to verify each other. Visually gathered spatial cues are substantiated by auditory temporal cues[78]; olfactory and tactile cues fill in the density, so that at every moment we are present in a cognitive reality that we co-create with our surroundings. We humans are hardwired to give priority to new and changing inputs. For this reason our peripheral vision is much more sensitive to motion than our gaze,[79] and our "peripheral hearing" as a perceptual field can be much larger than our immediate surroundings. Peripheral hearing is how we remain alert to changes in our surroundings without always needing to have our guard up. It is the field just beyond our cognitive map which we monitor only, allocating most of our attention on our immediate setting. When we construct our auditory maps, we create the "known field" that will contrast with any peripheral activity, helping us discriminate between the two.

We humans have an extensive array of cognitive tools to help construct and reconcile our acoustical maps, or "auditory scenes." These include ways of integrating simultaneous and sequential sounds; fusing temporal and spatial sound patterns; segregating overlapping sounds; stripping out coherent sounds from louder background noise; finding melodies and rhythms in soundfields of chaos, and identifying the most miniscule harmonic changes in the din of our environment.[80] We don't do this solely to enjoy music or understand speech; rather these capabilities allow us to maintain a perceptual grasp on where we are—for our own survival. (You might say that speech and music are pleasant artifacts of these perceptual tools.)

Of course we are not just dwelling in our auditory scene, we participate as well, and this is where speaking and singing—our intentional communication channels—enter the territory. In its most basic form our self generated sounds are declaratory mapping cues, stating "I am here and this is what is happening." This works in any proximity, from yelling across the playground to the whispers and breadth of making love. We "sing along" with our environment, modifying our own auditory maps and influencing the maps of others in a manner akin to the acoustic-spatial domains of stridulating insects, flocking birds, and schooling fish.

The sense of human participation is really evident in the forest singing of the Kaluli, whose bird language is formed by their relationship with nature. At first blush their jungle singing sounds more like exotic birds in the forest, undulating to the pulse of the cicadas, opening to acoustical spaces available to them as the jungle animals respond to their voices.[81] The incantations of the Apache fence-stringer mentioned before punctuate this participation, much in the manner that the Aborigine Australians sing their songlines across their ancient landscape, and a clear example of a cognitive sense of place and a conscious expression of that sense through sound.

The Aboriginal songlines are a mapping system that is quite literal, such that a person born under a songline or "Dreaming" of a particular totemic animal is responsible for keeping their map intact. They must help re-sing the song of the Dreaming whenever they come across the songline. This is geographically significant because the Songlines weave across the whole continent—many over 3,000 miles in length, and quite detailed and true.

Bruce Chatwin, in his book *"The Songlines"* tells about an incident he had while driving across the desert in a pick-up truck with a man of the Native Cat Dreaming. The man, who had been silent throughout the ride, at an instant started ripping off some garbled rapid-fire song, sounding like the wind rustling through the trees. They had come across his songline, but traveling at 25 m.p.h. along a songline that was meant to be sung at a walking pace of 3–4 m.p.h. Once he slowed the vehicle down to a stroll, the melody of the song revealed itself.[82]

For the Aborigines, their songlines—these songs—are the thin threads that bind the world together and appliqué all of us to the fabric of the universe. As one tribal elder remarked; "If we do not sing the songs the animals will go away. Then we will all die."

This song-mapping is not exclusive to the Aborigines. It seems that the Indigenous people of the Pacific Northwest—the Nootka, the Haida, the Kwakiutl, and the Bela Coola—also mapped and navigated by songs:

> Everything we ever knew about the movement of the sea was preserved in the verses of a song. For thousands of years we went where we wanted and came home safe, because of the song. ...There was a song for goin' to China, and a song for goin' to Japan, a song for the big island and a song for the smaller one. All she had to know was the song and she knew where she was. To get back, she just sang the song in reverse...[83]

Their songs invoke the memory of earlier journeys, but they also sing a running response to their surroundings: the features in these maps invoke expressions of their presence in the song,[84] perhaps in a manner akin to how low-frequency geographical sound-cues and dynamic weather fronts invoke the migration paths of certain birds,[85] and the chorusing soundfields of the cicadas and crickets define the perimeters of their domain. All of these beings have perceptual platforms seated in spatial and temporal priorities suited to their survival, and in a similar sense they are all viscerally tied into their worlds by way of sound.

The challenge of understanding communication at this level is posed by the fact that each animal species frames their perceptual platforms in manners that are both specifically expanded and generally limited by their form and behavior, their biological equipment, and their particular sense of time. The spatial and temporal priorities of bird and human navigations are large and long; the chorusing of crickets, the tight flocking of birds and the schooling of fish are more spontaneous and contained in a closer range; the high frequency sonar calls of bats and dolphins are instantaneous and in tight precision. These animals all have different temporal and spatial frameworks and priorities—but with their own tools they all "sing their surroundings."

If we return to the idea of "acoustic daylight" we find that some animals may be able to resolve the sound of their surroundings by way of "passive sonar." This is the ability to resolve detailed environmental information by listening to sounds alone (rather than actively producing a sound and listening to how it bounces back). Passive sonar is what the crickets may be doing by listening to each other's sound to determine the nature of their surroundings. If we kick this idea up into the ultrasonic realm of dolphins, and include the neural processing density that dolphins have available to them, it is quite possible that these animals may communicate to each other by way of "sonic holograms." If a dolphin can bounce a complex signal off of a fish and read the return signal as a finely detailed sonogram, it is likely that other dolphins in their immediate surroundings may also be able to "read" the sonograms generated by their kin.[86] From this point it is not a long reach to the idea that dolphins may also be able to project sonar images to each other, thus singing their imagination to their kin.[87]

While this may seem far-fetched, dolphins do have the right equipment for the task. Given that they can vocalize across a frequency band of 1–150 kHz, and have the ability to "stereophonate,"[88] they would possess the acoustical tools to craft detailed "images" in their surroundings. This communication would not be based on words, representations, or meaning, rather it would be a direct co-creation of the communication soundfield with the field and those within it. It is clear that dolphins (and porpoises) are able to communicate very complex information to each other. If they do so with "sonographic communication," it would explain why Dr. John Lilly—considered the Godfather of dolphin communication research, could not decipher the language of dolphins after 18 years of dense, systematic study; Lilly was looking for communication in representative or symbolic sounds, not in sonograms.[89] If dolphins can craft acoustical fields in their environment, it could be the most refined variation of communication, bringing sound into form by "singing their surroundings."

I began this chapter saturated in the sounds of a Spring dawn chorus, high up in an alpine hot-spring meadow. It is now Autumn. I am in my back yard this morning, listening to the sounds of migrating birds, which bear a significantly different texture than the sounds of spring. The birds have been traveling throughout the night. They come down to earth in the early morning to feed, rest, and dwell until the late afternoon thermals help them back up into their heavenly migration routes.

Many of these migrating birds travel in mixed groups rather than in flocks, because their genetic business is taken care of; thus the young dusky robins will fly with the mourning doves because of their similar size, not because of a need to bond with kin. This makes for a greater variety of birds in any setting because they are out of habitat, and not particularly territorial. They are also not singing their urgent breeding and territorial songs. Their purpose is scattered, and so their sounds are as well. The vocalizations are not produced from deep draughts of air through their syrinx, rather most birds are just clicking and plittering away. Their shallow sounds seem to emanate from their beaks—sounds that are called "chip notes," used as gentle reminders of their presence. Even the song sparrows, who embroider the Spring mornings with fabulous sonic fanfares are now only singing a few plaintive descending tones.

What I also notice are the various sounds of their wings: A small finch harasses a thrush with a hot pursuit. As the finch closes in, the thrush claps her wings hard— "flap! Flap! FLAP! ...enough! dammit!..." landing into the low branches of a fir tree. A flight volley of two doves dart and weave though the treetops to the synchronized patting of their wingbeats; wrens and bush-tits flitter about the fading yellow leaves of a mountain ash, gleaning spiders and mites for their breakfast. A flycatcher flings himself off of his high perch, then flips back on with a wing and a wave.

I am taking this in; watching an Allen's hummingbird search for mites in the needles of a redwood, hovering delicately; not with the taught energy "hum" of a spring nectar diet, but with a light flutter and relaxed ease of a runner pacing himself. He flies over me and gives his wings a light "snap!" as he clears my head. In the next few weeks he will fly a thousand miles.

Notes

1. Stocker, M. (1995). Museum sound design. *Museum International, 185*, 25 UNESCO.

Notes for Chapter 1

1. Pratarelli, M. E., & Steitz, B. J. (1995, April). Effects of gender on perception of spatial illusions. *Perceptual and Motor Skills, 80*(2), 625–626. Also Lewald, J. (2004, March). Gender-specific hemispheric asymmetry in auditory space perception. *Brain Research: Cognitive Brain Research, 19*(1), 92–99.
2. Rosen, Plester, El-Mofty, & Satti. (1962). *Annual of Otology, Rhinology and Laryngology.*
3. Marks, I. M. (1969). *Fears and phobias.* Academic Press.
4. Wachs, T., & Gruen, G. (1982). *Early experience and human development.* New York, NY: Plenum.
5. Deliege, I., & Sloboda, J. (Eds.) (1996). *Musical beginnings.* Oxford University Press.
6. All airborne sound pressure levels in decibels (dB SPL) used in this book are referenced to 20 μPa unless otherwise noted. This convention is consistent with referencing auditory SPL to the lowest threshold of human sound perception.
7. Morris, D. (1968). *The naked Ape* (pp. 106–109). New York: McGraw-Hill Book Company.
8. Satt B. PhD. *Sounds in the womb: What do babies hear before birth?* See also: Dr. Thurman, L., & Langness, A. P. (1986). *Heartsongs: A guide to active pre-birth and infant parenting through language and singing.* Music Study Services.
9. The late Tabla Master Alla Rakha would play his rhythms on the taught belly of his pregnant wife. His son, Zakir Hussein, took in these early lessons well to become one of the world's eminent Tabla players.
10. A "Big Wheel" is a blow-molded polyethylene tricycle. Due to all of the hollow-cast resonant cavities constituting its form, the cavities serve as "sound boxes," amplifying the wheel-to-surface contact throughout its body, almost "thundering" when ridden across any rough surface.
11. Truax, B. (1984). *Acoustic communication* (p. 113). Norwood, NJ: Ablex Publishing Corp.
12. On habituation, see: Doelle, L. L. (1972). *Environmental acoustics* (p. 137). McGraw-Hill.
13. Also on habituation, see: Truax, B. (1984). *Acoustic communication* (p. 90). Norwood, NJ: Ablex Publishing Corp.
14. Ibid. pp. 24–25.

M. Stocker, *Hear Where We Are: Sound, Ecology, and Sense of Place,*
DOI 10.1007/978-1-4614-7285-8, © Michael Stocker 2013

15. Murray Schafer, R. (1994). *The soundscape: Our sonic environment and the tuning of the world* (pp. 74–75). Rochester, VT: Destiny Books.
16. William M. C. Lam (1977). *Perception and lighting as formgivers for architecture* (p. 21). New York: McGraw Hill Book Company.
17. Ibid. p. 14.
18. Árvay, A. (1969, December). *Effect of noise during pregnancy upon fetal viability and development.* Presented at the International Symposium on the Extra-Auditory Physiological Effects of Audible Sound held in Boston, MA (Published in "Physiological Effects of Noise" 1970, Plenum Press, NY).
19. Jensen, G. (1991). Physiological effects of noise. In C. M. Harris (Ed.) *Handbook of acoustical measurement and noise control* (Chap. 25). McGraw Hill.
20. Berglund, B., Lindvall, T., & Schwela, D. H. (Eds.) (1999). *Adverse health effects of noise.* World Health Organization Guidelines for Community Noise.
21. Wachs, T., & Gruen, G. (1982). *Early experience and human development.* New York, NY: Plenum.
22. Bronzaft, A. L. (1981). The effect of a noise abatement program on reading ability. *Journal of Environmental Psychology, 1*, 215–222.
23. Truax, B. (1994). *Acoustic communication* (Ch. 6: Noise and the urban soundscape). Norwood, NJ: Ablex Publishing Corp. See also. p. 11, 20, and 52.
24. Plato "Republic" Book III v. 398–402. Also *Musical training is a more potent instrument than any other, because rhythm and harmony find their way into the inward places of the soul.* (F. M. Cornford Trans.) Oxford University Press. See also Aristotle "Politica" Book VIII Ch. 5 on music education.
25. For Cages writing and ideas, see *Silence: Lectures and writings.* MIT Press (1966) and *A year from Monday.* Wesleyan University Press (1967).
26. Stephen Coleridge quoted in Betjeman, J. (1969). *Victorian and Edwardian London* (pp. ix–xi). London: B.T. Batsford.
27. Yi-Fu Tuan. (1979). *Landscapes of fear* (p. 153). New York: Pantheon Books.
28. Juvenal. (circa 75–100 A.C.E). *Against the City of Rome: The third Satire* (lines 232–238).
29. Saunders, N. K. (Trans.). (1971). *The epic of Gilgamesh* (Sumerian poem, c. 3000 B.C.E.). Middlesex: Harmondsworth. Quote found in Schaefer, R. M. *Tuning of the world* (p. 105).
30. Rossing, T. D. (1989). *The science of sound* (2nd ed., p. 625). Addison Wesley.
31. Fitch, T., & Kramer, G. (1994). Sonifying the body electric: Superiority of an auditory over a visual display in a complex multivariate system. In G. Kramer (Ed.), *Auditory display: "Sonification, audification and auditory interfaces: Proceedings from the 2nd International Conference on Auditory Display".* Reading, MA: Addison Wesley.
32. Nixon, C., Anderson, T., Morris, L. J., McCavitt, A., McKinley, R., Yeager, D. G., & McDaniel, M. P. (1998). Female voice communications in high level aircraft cockpit noises: Part I: 675–683-Part II 1087–1094. *Aviation, Space and Environmental Medicine, 69.*
33. Merker, H. (1994). *Listening: Ways of hearing in a silent world* (p. 154). HarperCollins Publishers.
34. Dr. Salk, L. (1960). The effects of the normal heartbeat sound on the behavior of the newborn infant: Implications for mental health. *World Mental Health, 12*, 1–8.
35. Satt, B. PhD. *Sounds in the womb: What do babies hear before birth?*
36. Weisburd S. (1992, April). Grandpals, link of ages: linking nursing homes and day-care centers. *Health, 6*(1) Copyright 1992 Hippocrates Inc.
37. From "Vital Signs" *In Health*, May/June 1990.
38. Human sound perception has a dynamic range of 120 dB, meanwhile theater sound production has a much more limited dynamic range of 60–75 dB. In order to simulate the range of human perception in the theater—from whispers and leaf rustles to explosions and screams, the dynamic range needs to be "compressed" so that quite signals are produced much louder and loud signals are produced much quieter than what they would be in their real form.
39. Personal conversation.

40. Produced by Reggio, G. (1981). *Koyaanisqatsi: Life out of balance.* Institute for Regional Education.
41. See (1994) Schizophonia. In B. Truax (Ed.) *Acoustic communication* (p. 120). Norwood, NJ: Ablex Publishing Corp. Also Murray Schafer, R. (1994) p. 90.
42. DeLillo, D. (1984). *White noise.* New York: Viking Penguin Press.
43. Helmi Järvilouma interview with Murray Shafer, R. (1994). *Soundscapes; essays on Vroom and Moo* (p. 117). Department of Folk Tradition, Institute of Rhythm Music.
44. Lejeng Kusin, an older Penan woman, Ubong River, May 1993 from Davis, W., Mackenzie, I., & Kennedy, S. (1995). *Nomads of the dawn: The Penan of the Borneo Rain Forest* (p. 22). Pomegranate Books.
45. Pöyskö, M. (1994). The blessed noise and little moo: Aspects of soundscape in cowsheds. In *Soundscapes; essays on Vroom and Moo* (p. 83). Department of Folk Tradition, Institute of Rhythm Music (1994).

Notes for Chapter 2

1. The Bible, John 1:1.
2. Kramer, S. N., & Wolkstein, D. (1983). *Inanna: Queen of heaven and earth* (p. 123). New York: Harper & Row.
3. The Koran. The question of the omnipotence of Allah is laid to rest by the ultimate faithful acceptance of knowing that all Allah has to do is proclaim "Be" to manifest the Universe. Among other places this is expressed in Al-'Imrān 3:48 wherein the angels explain to the Virgin Mary how she could give birth without having been touched by a man, and in Al-An'am (Cattle) 6:73 in a general rallying of the faithful: "On the day He says 'Be' it shall be."
4. Chatwin, B. (1987). *The songlines.* Penguin explores the songlines in a biographical form through the authors relationships with Aboriginal people.
5. Hartshorne, C. (1973). *Born to sing* (p. 54). University of Indiana Press. The author postulates that birdsong serves a deeper purpose than just courtship and territorial concerns, suggesting that birds may actually be able to convey "feelings" about that which they sing.(!)
6. Plato "Phaedrus" 259.
7. Michael Pollan "Botany of Desire" suggests that a plants "usefulness" to humans in their beauty or nutritional value is the plant's method of inducing humans to cultivate the plant— ostensibly serving the plant's need to procreate.
8. Meeuse, B., & Morris, S. (1984). *The sex life of flowers* (Facts on File books, Chap. 5).
9. Benzon, W. (2001). *Beethoven's Anvil: Music mind and culture.* Basic Books. Cognitive scientist Benzon's book looks at the role of music in culture. A dominant theme in the book is that community bonding through singing, music and dance serve a biological purpose of binding the individual into the fabric of their community.
10. Hartshorne, C. (1973). *Born to sing* (p. 51). University of Indiana Press. Examines the sound play and sound engagement with objects demonstrated by young birds. "More than all creatures save man, birds play with sounds, not only sound produced directly by their own organs, but even those resulting in handling of objects" exclaiming in response to the sounds of the objects in play.
11. Aitchison, J. (1996). *The seeds of speech: Language origin and evolution* (p. 13). Cambridge University Press. The are many plausible theories on the origin of language in humans. Most of them are probably correct if taken along with the array of the other theories. Aitchison suggests that "language need not have been started with single words, it could have begun with melodies."
12. The Bible. Genesis 2:18–19 also 1:28–30.
13. The Koran "Al-Baqarah" (The Cow) 2:29–34. The angels ask: "Will You put there one that will do evil and shed blood, when we have for so long sung Your praises and sanctified Your

name?" In this exchange, God has Adam tell the Angels their names, then instructs them to "Prostrate yourselves before Adam." They all do—except Satan, who becomes the first unbeliever.

14. Wallis Budge, E. A. (1904). *The Gods of the Egyptians* (Vol. 1, p. 109). (Dover Edition 1969). also see: Chatwin, B. (1987). *The songlines* (p. 271). Penguin Books: New York.

15. Ibid. p. lvii–lxix.

16. Wallis Budge, E. A. (1904). *The Gods of the Egyptians* (Vol. 1, p. 301). (Dover Edition 1969).

17. This metaphor remains today in among other places, the traditional wedding ceremony of the Mormons. In passing through the veil into the bond of marriage, the wife reveals her celestial name to her husband, submitting to his authority. Prior to this moment, only she and her elder know this name. Once she has submitted her celestial name to her husband, he must recall it in the hereafter before she is allowed admittance to Heaven.

18. Wallis Budge, E. A. (1895). *The Egyptian book of the dead: The Papyrus of Ani.* (Dover Edition 1965). List of Chapters: xxxiii on "Giving Mouth" p. 307, Chapter XXV "Causing a man to remember his name." p. 265–269 on the details of "opening the mouth." In a more metaphorical context, immortality depended on the "...heirs who were children who could pronounce their names..." One of the perks of being a Scribe included making "...heirs for themselves of the writings..., [which] cause him to be remembered in the mouth of the storyteller." — Simpson, W. K. (Ed.). (1973). Papyrus of Chester Beatty IV. In *Literature of Ancient Egypt.* Yale University Press.

19. Wallis Budge, E. A. (1904). *The Gods of the Egyptians* (Vol. 1, p. 301). (Dover Edition 1969). Osiris' self creation foreshadows the Christian voicing of the concept of "In the beginning was the word..." Before telling us what the "Word" was, the Gospel of John tell us that the "...Word was with God..." he then tells us that the "...Word was God."

20. The Bible "The Revelation of St. John" 2:17. wherein Christ dispenses with these names as a reward to believers in the "Seven Churches of Asia".

21. Abram, D. (1996). *The spell of the sensuous* (p. 80). Vintage. Chapter 3 "The Flesh of Language" sensuously and articulately expands on this theme.

22. Foucault, M. (1970). *The order of things: An archeology of the human sciences.* Random House, explores the history of cognitive relationships between words and things in western culture, particularly examining the transition between the Classical age and the Modern age. This examination traces the transformation of language from words as representative of things to words as a vehicle for thought. An illuminating discussion therein looks at all languages in terms of "inflection" and "internal variations" stating that "In order to determine the primary and quite simple elements of a language, general grammar was obliged to work backwards to that imaginary point of contact where sound, as not yet verbal, was in some sort of contact within the vital energy of representation." p. 288 of the 1994 Vintage Edition.

23. The Bible. Genesis 27:27–28.

24. Chief Marie Smith Jones May 14, 1918–January 21, 2008.

25. Diné and Eyak are both Athabaskan dialects—one of the five major indigenous language families found in the western hemisphere. The Diné (also known as "Navajo") is commonly spoken in the Arizona and New Mexico area of the U.S. with over two million speakers.

26. William Shakespeare *Hamlet, Prince of Denmark* Act III, Scene II.

27. Aristotle *Rhetoric* Book III (Chap. 8).

28. Watling, E. F. (1947). Introduction to "Sophocles: The Theban Plays" (Trans.) (p. 20). (From the 1969 Penguin Books edition).

29. Vovolis, T., & Zamboulakis, G. (1994–2012). The acoustical mask of Greek tragedy. *Didaskalia.* ISSN 1321-4853. See: http://www.didaskalia.net/issues/vol7no1/vovolis_zamboulakis.html

30. Sanders, B. (1994). *A is for Ox* (p. 56). Pantheon Books.

31. Augustine. (~386 C.E.). *Confessions* Book VI:3 attempts to guess why his teacher, the Bishop of Milan, reads silently. This is probably the first mention of silent reading in existing western literature.

32. Weghorst, A. A. (1997, May). *The Question of Literacy in Miguel Cervantes' Don Quixote*. The Trinity Papers. Trinity College.

33. Sanders, B. (1994). *A is for Ox* (p. 67–68). Pantheon Books.

34. Ibid. p. 3–4.

35. Reynolds, C., Smith, M., & Woodman, M. (2002, December). Longer than a telephone wire—Voice firewalls to counter ubiquitous lie detection. *Journal of the Institute of Technology at Blanchardstown, Dublin* Issue 6.

36. This account comes from a personal conversation with a friend who lived communally with Pearls in the 1960s.

37. Nicastro, N. (2002, June). *Acoustic correlates of human response to domestic cat (felis catus) vocalizations*. Delivered at the 143rd ASA meeting 2002, Pittsburgh PA. Nicastro does not go so far as to call these paralinguistic sounds "vocabulary," as the sounds do not represent anything specific, rather they are sounds of persuasion that induced predictable responses in humans.

38. Price, P. (1983). *Bells and man* (p. 271). Oxford University Press. The largest manufactured bell in existence.

39. Sexsmith, P. (2001). *Portrait of a jingle dress dancer*. Guide to Indian Country, Aboriginal Multi Media Society, Saskatchewan, Canada.

40. Sloan, E. (1966). *The sound of bells* (p. 21). Doubleday and Co.

41. Price, P. (1983). *Bells and man* (p. 1). Oxford University Press.

42. Schafer, M. (1997, 1994). *The soundscape: Our sonic environment and tuning of the world* (p. 53–54). Destiny Books.

43. Price, P. (1983). *Bells and man* (p. 79). Oxford University Press.

44. Stalley, R. (1999). *Early medieval architecture* (p. 123). Oxford University Press.

45. Leroux-Dhuys, J.-F. (1998). *Cistercian Abbeys: History and architecture* (p. 56), Könemann Illustrates the monk's time schedule in its complexity. The public schedule intersected this at many places but was not entirely ruled by all of the offices.

46. Gimple, J. (1961, 1984). *The cathedral builders* (p. 1). Harper Colophon.

47. Corbin, A. (1994). *Village bells: Sound and meaning in the nineteenth century French countryside* (p. 110–118). Columbia University Press. Precise time keeping was introduced into French law in 1891, though it was not universally practiced up into the twentieth century. In provincial areas the "Angelus" still rings to the rhythm of the church.

48. The Christian use of Bells to call the faithful, or to work miracles, begins seeping into the legacy at the late part of the Roman Empire, when Christianity was tolerated—then finally adopted. Prior to that time, using bells to call the faithful to a "prohibited religious service" would have been counterproductive. For early Christian use of bells, see Price, P. (1983). *Bells and Man* (The introduction of bells into the church, Chap. 4). Oxford University Press.

49. Corbin, A. (1994). *Village bells: Sound and meaning in the nineteenth century French countryside* (p. 90, 142–145). Columbia University Press.

50. Illich, I. (2000). *The loudspeaker on the tower*. Bremen.

51. Corbi, A. (1994). *Village bells: Sound and meaning in the nineteenth century French countryside*. Columbia University Press examines this in Proustian detail.

52. Price, P. (1983). *Bells and man* (p. 233). Oxford University Press. Photo from Library Archives Canada MUS 133, Accession 198131, Volume 239, (box barcode 201523415) folder 768A.

53. Henry M. Stanley "Through a Dark Continent" Dr. David Livingstone "The Life and African Exploration of Dr. David Livingstone: Comprising All His Extensive Travels and Discoveries As Detailed in His Diary, Reports, and Letters, Including His Famous Last Journals".

54. Bebey, F. (1969). *African music: A people's art* (p. 98). Lawrence Hill & Co. Radio Ghana's station identification "Ghana Listen! Ghana Listen!" can be heard on short wave radio by "ham" operators worldwide. The author recorded some short wave radio log drum "call letters" in the early 1990s, confirming this practice as late as that date.

55. Steintorf, L. A. (1950). *White witch doctor* (p. 30–31). Westminster Press.

56. See also Schultes, R. E., & Raffauf, R. F. (1992). *Vine of the soul: Medicine men, their plants and rituals in the Colombian Amazon* (p. 176). Synergetic Press: Arizona. "Witotos signaling for a tribal ceremony…can be heard 10 or 12 miles away, sometimes even farther if the sound travels along a river and not through the forest."

57. The development of the minaret can be traced to watchtowers, lighthouses, signal relay towers, and beacon towers for travelers—all structures with a visual legacy. See: Sims, E. (1984). Markets and Caravanserais. In G. Mitchell (Ed.), *Architecture of the Islamic World* (p. 99). Thames and Hudson.

58. Steinsaltz, R. A. (2000). *A guide to Jewish prayer* (p. 185–188). Israel Institute for Talmudic Publications.

59. The Bible. Joshua 6:1–20.

60. Keizer, G. (2001, March) Sound and Fury: The politics of noise in a loud society (p. 40). Essay in *Harpers Magazine*, so eloquently put it: "Your ear is my hole."

61. Abram, D. (1996). *The spell of the sensuous* (p. 256). Vintage. "…for many oral, indigenous peoples, the boundaries enacted by their languages are more like permeable membranes binding the peoples to their particular terrains, rather than barriers walling them off from the land. By affirming that the other animals have their own languages, and that even the rustling of the leaves in an oak tree or aspen grove is itself a kind of voice, oral peoples bind their senses to the shifting sound and gestures of the local earth,…"

62. Somé, M. P. (1999). *The healing wisdom of Africa* (p. 93). Jeremy P. Tarcher/Putnam.

63. Smale, A. (1997, January 25). *Whistle is Serbian rebel weapon.* Associate Press.

64. 410 U.S. 113 U.S. Supreme Court. "Roe et al. V. Wade, Dallas County District Attorney" January 1973 Case that secured rights for American women to obtain an abortion on demand.

65. At the time of this writing in 2007, Representative Pelosi is the Democratic Majority leader and Speaker of the House of Representatives.

66. Orwell, G. (1941/2003). *Animal farm* (p. 25). Plenum/Harcourt Brace.

67. This strategy was more than imagined in a 1980s action of dissent in Germany protesting a increase in NATO tactical nuclear weapons in their country. Silent protesters would stand mute, *en masse* indicating that there was "absolutely nothing to say about nuclear destruction." See Illich, I. The right to dignified silence. In *In the mirror of the past* (p. 27). Marion Boyars.

68. Plato "Laws" Book V [741] (B. Jowett, Trans.). Citizens would be landholders. The framing of scale around the auditory reach of an orators voice is suggested by Murray Schafer, R. (1977, 1994). *The soundscape: Our sonic environment and tuning of the world* (p. 215). Destiny Books.

69. Illich, I. (1983, Winter). Silence is a commons. *CoEvolution Quarterly.* Illich's writings were an ongoing process until his death in Dec. 2002. This quote comes a few paragraphs from the end.

70. Hitler, A. (1938) *Manual of German Radio.*

71. Speer, A. (1970). *Inside the third Reich* (p. 70–71). MacMillan. Talks about reading the inspiring speeches of der Fürer after the fall of the Reich. He was disappointed to find them "empty, shallow, and useless…incomprehensible that these tirades should once have impressed me so profoundly."

72. Ibid. p. 32

73. Thompson, E. (2002). *The soundscape of modernity* (p. 233). MIT Press. Her chapter on Electro-acoustics and Modern Sound 1900–1933 is a great read on how engineers and technologists had to grope through this cognitive shift while producing the "talkies."

74. Murray Schafer, R. (1977, 1994). *The soundscape: Our sonic environment and tuning of the world* (p. 88). Destiny Books.

75. Mander, J. (1978). *Four arguments for the elimination of television, The replacement of human images by television* (p. 240–260). Quill Books.

76. Carsey, M., & Werner, T. (2000). Father of broadcast David Sarnoff. *Time Magazine.* "Now we add sight to sound. It is with a feeling of humbleness that I come to this moment of announcing the birth in this country of a new art so important in its implications that it is

bound to affect all society. It is an art which shines like a torch of hope in the troubled world. It is a creative force which we must learn to utilize for the benefit of all mankind. This miracle of engineering skill which one day will bring the world to the home also brings a new American industry to serve man's material welfare ... [Television] will become an important factor in American economic life."

77. An adage among the Society of Motion Picture and Television Engineers is "Television without picture is just radio; television without sound is 'technical difficulties.'"

78. Propaganda newsreels produced by "Time. Inc." played at movie house matinees through and after WWII.

79. Harding, J. R. (1973/4). The Bull-Roarer in history and antiquity. *Journal of The African Music Society, 5*(3), 42.

80. Larche, L. (2002). Pipers play to a different tune. *Fort Polk Guardian, 29*(11).

81. Price, P. (1983). *Bells and man* (p. 50). Oxford University Press. "Their feelings when they heard the metal jangling on a foe would be comparable to those of a people without defense of aircraft today on hearing the buzz of an enemy airplane." p. 74

82. Ear witness comment. The Buzz Bomb sound signature was produced by a pulsed-jet engine that propelled this sub-sonic missile.

83. *Arrow-Odd: A medieval novel.* (P. Edwards, & H. Palsson, Trans.) (p. 40). (1970). New York University Press/University of London Press. The Berserkers were thought to be a cult of Odin, their frenzy an ecstatic state either self induced or enhanced by the entheogenic mushroom *Amanita Muscaria*.

84. Brigadier Gen. David Grange, U.S. Army (retired): "I mean the psychological effect to the kinetic effect is probably three to one....So the psychological impact is more important, actually, than the kinetic impact." Interview with CNN news anchor Paula Zahn about the MOAB ordinance. Aired March 12, 2003.

85. Ross, C. D. (1999, Winter). Outdoor sound propagation in the U.S. Civil War. *Echoes, The newsletter of The Acoustical Society of America, 9*(1).

86. Odd, but a proud claim to fame for this Dutch coastal town.

87. The Bible: Judges 12:5–6

88. See Maalouf, A. (1985). *The crusades through Arab eyes*. Schoken Books: New York, for a besieged perspective on medieval warfare, and Tripoli's 2000 days of siege. Vitruvius "Ten Books on Architecture" Dover Edition. Book 1 (Chap. I:8) and Book 10 (Chap. XII) mentions the need to tune the pitch of catapult ropes to assure accuracy—applied acoustics in action. In Book 10 (Chap. XIII–XVI) talks about siege machines and defense.

89. Herodotus. Historical event in sixth Century BCE Barca (Chap. 4:200).

90. Vitruvius Book 10 (Chap. XVI:10). Dover Edition.

91. Illich, I. (2000). *The loudspeaker on the tower*. Bremen.

92. Corbin, A. (1994). *Village bells: Sound and meaning in the nineteenth century French countryside* (p. 5). Columbia University Press.

93. Altmann, J. (1999, May). *Acoustic weapons—A prospective assessment: sources, propagation, and effects of strong sound*. Cornell University Peace Studies Program Occasional paper #22 While loss of bowel control is theoretically possible at high levels of infrasound, there is no strong evidence of reliably producing this effect. Annoyance thresholds are above 130 dB at 100 hz–50 Hz. Rupture of the ear drum is ~186–189 dB.

94. From Atlas/Soundolier catalog.

95. Anticaglia, J. R. (1970). Extra-Auditory Effects of sound on the Special Senses. In B. L. Welch, & A. S. Welch (Ed.). *Physiological effects of noise*. Plenum Press: New York.

96. 125 dB SPL re: 20 µPa (Decibels in an airborne environment are most commonly referred to relative to 20 µPa, or "0 dB SL"—the apparent threshold of human hearing.)

97. Reeves, R., Hofman, R., Silber, G., & Wilkinsen, D. Acoustic Deterrence of Harmful Marine Mammal-Fishery Interactions. 1996 workshop proceedings. NOAA, NMFS

98. The alien program material probably had something to do with its effectiveness; Mr. Noriega was a known opera buff—which had some bearing on the material selection. David Koresh's

Branch Dividians endured Nancy Sinatra's "Boots" at exceedingly high volumes during the assault of their Waco, Texas commune.

99. Muller, J. (2002). Sound and fury: sonic bullets to be acoustic weapon of the future. *ABC News report*. A typically sensationalistic presentation about "Acoustic Weapons" strung over a web of factoids: "The operator chooses one of many annoying sounds in the computer—in this case, the high pitched wail of a baby, played backwards—and aims it at us. At 110 decibels, we were forced to walk out of the beam's path, our ears ringing. Had we stayed longer, Norris said our skulls would literally start to vibrate."

100. 1st. Lt. Rob Miller interview with Jack Hitt "In Country: What life is like for American Soldiers in Iraq" from "This American Life" WBEZ 1/07/05

101. American Technology Corp. San Diego, CA has developed an array of "non-lethal" acoustic devices including the "Long Range Acoustic Device" (LRAD), High Intensity Directed Acoustic (HIDA) devices, and other novel sound equipment. Since the 2000 attack on the USS Cole, LRAD devices have been used by US Navy ships and other shore boats to ward off pirates and other attackers. April 2006, "Acoustics Today," publication of the Acoustic Society of America.

102. *Police ready sound weapon for GOP Convention.* (2004, August 19). Associated Press.

103. Active sonar involves sending out an acoustic signal, passive sonar is just listening for acoustical information without first sending a signal out.

104. Stocker, M. (2002). Ocean bio-acoustics and noise pollution: Fish, mollusks and other sea animals' use of sound and the impact of anthropogenic noise in the marine acoustic environment. *Soundscape Journal of Acoustic Ecology, 3*(2)/*4*(1).

105. Decibels in a marine environment are most commonly referred to relative to 1 μPa by convention. The numerical difference between airborne sound levels and marine sound levels is ~61 dB, so 215 "Ocean Decibels" (ref: 1 μPa) is the equivalent energy level as 155 dB (20 μPa) in air, expressing a numerical difference of 26 dB (20log 1 μPa/20 μPa) plus a property difference of 35 dB (10log 1/3500)—water is ~3,500 times denser than air. For physical reference, 125 dB broad band noise in air is the human pain threshold. With broad band noise, 155 dB in air is just shy of where the human eardrum will implode, and momentary exposure will cause irreparable ear damage.

106. Department of the Navy, Chief of Naval Operations. (2001). *Final Overseas Environmental Impact Statement and Environmental Impact Statement for Surveillance Towed Array Sensor System Low Frequency Active (SURTASS/LFA) Sonar System.* Vol. 1. section 4.3. p. 4.

107. Berent, J.-E. (1987, 1991). *The world is sound Nada Brahma* (p. 27). Destiny Books.

108. Homer *Odyssey* Book IV:233 "Moreover, every one in the whole country is a skilled physician, for they are of the race of Pæeon." (S. Butler Trans.)

109. The Bible. Mark 7:34.

110. Abrams, D. (1996). *The spell of the sensuous: Perception and language in a More-Than-Human World* "The traditional or tribal shaman, I came to discern, acts as an intermediary between the human community and the larger ecological field..." (p. 7). Vintage Books. Abrams indicates that the shaman may exploit a patients fears of the "spirit world" to dispel a patients belief in their illness, jarring them back to a well state.

111. Eliade, M. Shamanism: Archaic techniques of ecstasy. *Bollingen Series 76.* Princeton University Press.

112. Goldman, J. S. (1991). Sonic entrainment. In *Music: Physician for times to come* (p. 217–233). Quest Books. Anthology of music and healing literature assembled by Don Campbell.

113. Benzon, W. (2001). *Beethoven's Anvil: Music in mind and culture* (p. 6). Basic Books.

114. McNeill, W. H. (1995). *Keeping together in time: dance and drill in human history.* Harvard University Press; Cambridge, MA.

115. von Muggenthaler, E. (2001). The felid purr: A healing mechanism? *The Journal of the Acoustical Society of America, 110*, 2666.

116. Rubin, C., & McLeod, K. (1994). Promotion of bony ingrowth by frequency specific, low amplitude mechanical strains. *Clinical Orthopedics and Related Research, 289*, 165–174.

117. Observations made at the Seattle Aquarium in their captive Puffin program.
118. Cochrane, A., & Callen, K. (1992). *Dolphins and their power to heal* (p. 28). Healing Arts Press.
119. See: Dobbs, H. (2000). *Dolphin healing: The science and magic of dolphins.* Piatkus Books
120. Nathanson, D. E. Ph.D. (1996, September). *Dolphin human therapy and research.* Proceedings from the Second Annual International Symposium on Dolphin-Assisted Therapy. (www. aquathought.com/idatra/symposium/96/symposium.html). The position of Nathanson (and Lindblad, below) is that this work is therapeutic by virtue of the playful social interaction between humans and other intelligent and sentient beings, not specifically as a result of the effects of sound of dolphin vocalization.
121. Lindeblad, S. K. *The effect of a unique stimulus (swimming with dolphins) on the communication between parents and their children with disabilities.* ibid.
122. Cole, D. M. *Electroencephalographic results of human—Dolphin interaction: A sonophoresis model.* ibid.
123. Atwater, F. H. *Complementary concepts on the effects of sound on consciousness.* ibid.
124. Sun, J., & Hynynen, K. (1999, April). The potential of trans-skull ultrasound therapy and surgery using the maximum available skull surface area. *The Journal of the Acoustical Society of America, 105,* 2519.
125. Capra, F. (1988). *Uncommon wisdom: Conversations with remarkable people* (p. 121), Bantam Books, conversation with Stanislav Grof on complimentary modes of consciousness.
126. Physical "resonance" of the body or the "ethereal body" is a dominant theme throughout the myriad books on sound and healing.

Notes for Chapter 3

1. Herman Helmholtz (1862). *On the sensations of tone* (p. 7) (from the Dover English Edition, 1954). Herbert Helmholtz was perhaps one of the last great generalists of the nineteenth century. His contributions to physiology, physics, astronomy, medicine and philosophy of education were all benchmarks, offered at the twilight of a "time in which a full synthetic view of nature was still possible." His magnum Opus "On the Sensations of Tone" continues to be the fundamental tome on which the exploration of acoustic physiology still stands.
2. Boyle, R. (1660). *New experiments physico-mechanical: Touching the spring of the air and its effects.*
3. Democritus, from a quote in Hunt, F. V. (1978). *Origins in acoustics* (p. 24). Yale University Press.
4. Arcytas fragment translated in Freeman, K. (1948). *Ancilla to the pre-socratic philosophers* (p. 79). Harvard University Press.
5. Plato (c. 428–348 B.C.E.) *"Timaeus" from "The dialogs of Plato"* (p. 67) (B. Jowett, Trans.). Oxford University Press.
6. Saeltzer, A. (1892). *A treatise on acoustics in connection with ventilation both ancient and modern.* D.Van Nostrand. Quoted in Rachel E. Dangermond essay "History" in Charles M. Salter Associates, Inc. "Acoustics" (p. 21) 1998 pub. William Stout Publishers.
7. See: *CRC Handbook of chemistry and physics* (61st ed., p. F-104). CRC Press.
8. Ibid. p. F-96. For gases and liquids, elasticity is expressed in terms of compressibility or the reciprocal of Bulk modulus—the modulus of volume elasticity.
9. I use the "pool of water" analogy with the caveat that it is not universally accurate. There are other factors that affect wave motion in water, but it does serve as a good visual tool.
10. Eshbach, O. W. (1954). *Handbook of engineering fundamentals* (pp. 9–37). John E. Wiley. An old text, but this aspect of physics hasn't changed since publication of this engineering classic.
11. (1979) *CRC Handbook of tables for applied engineering science* (Table 1–44. p. 90 and Table 7–43 p. 687). CRC Press.

12. O'Connell-Rodwell, C. E., Arnason, B. T., & Hart, L. A. (2000, December). Seismic properties of Asian elephant (Elephas maximus) vocalizations and locomotion. *Journal of the American Statistical Association, 108*(6), 3066.
13. For a delightful romp through Wallace Sabine's life and the profound impact his work had on the architecture of the twentieth and current centuries see: Thompson, E. (2002). *The Soundscape of Modernity: Architectural Acoustics and the Culture of Listening in America, 1900–1933* (pp. 33–113). Massachusetts Institute of Technology Press.
14. Rossing, T. D. (1990). *The science of sound* (p. 467). Addison-Wesley Publishing. A 'metric sabin' of a square meter of open window.
15. The unit measure of "Cycles per Second" is called "Hertz" in the parlance of engineers and physicists, honoring Heinrich Rudolf Hertz, who discovered radio waves. "Hz" is the abbreviation for "Hertz" and is used throughout this book for "Cycles per Second." Kilohertz is 1,000 Hz and is abbreviated as kHz following this convention.
16. A. S. Bregman's (1990) book *Auditory scene analysis* MIT Press provides an extremely comprehensive perspective on how humans construct "auditory scenes" of their surroundings, that even while unseen, include the perceptual integration of all that we hear in a given setting— from insects to motorboats, to our companions speaking to someone else across the room.
17. This time frame is the reciprocal of ~2,000 Hz—with a wavelength of about 7″.
18. Helmholtz, H. (1862). *On the sensations of tone* (From the Dover English Edition, 1954, p. 7).
19. Collier, J. L. (1989). *Benny Goodman and the swing era* (pp. 239–241). New York: Oxford University Press. After a 1936 New York "jam session," Goodman was so impressed by Lester Young's clarinet playing (on an inferior instrument) that he handed him his own clarinet.
20. See: Aristotle. *De Anima Book II* (418) (A. J. Smith, Trans.) and *On sense and sensible* (437) (J. I. Beare, Trans.) Plato. *Timaeus* [67] (B. Jowett, Trans.).
21. Kepler, J. (1618). *De Harmonica Mundi.*
22. von Békésy, G. (1960). *Experiments in hearing* (pp. 485–534). McGraw Hill.
23. Helmholtz, H. (1862). *On the sensations of tone* (pp. 137–138) (From the Dover English Edition, 1954). For a summary of these distinctions see: Buser, P., & Imbert, M. (1992). *Audition* (pp. 184–192). MIT Press.
24. Einstein, A. (2000) *The expanded quotable Einstein* (A. Calaprice (Ed.)) Princeton University Press.
25. Palmer, J. B. (1972). *Anatomy for speech and hearing* (2nd ed.). Harper & Row.
26. Buser, P., & Imbert, M. (1992). *Audition* (pp 110–112). MIT Press.
27. Ibid. pp. 148–150.
28. Handel, S. (1989). *Listening: An introduction to the perception of auditory events* (p. 64). MIT Press.
29. ~3.0 g/cm^3 for tooth enamel, ~2.5 g/cm^3 for petrous bone, ~1.95 g/cm^3 for hip socket bone.
30. Helmholtz, H. (1862). *On the sensations of tone* (p. 136) (From the Dover English Edition, 1954).
31. Lewis, E. R., Leverenz, E. L., & Bialek, W. (1985). *The vertebrate inner ear* (pp. 30–46). CRC Press.
32. Ibid. p. 7, pp. 220–222.
33. Ibid. pp. 5–6.
34. Yin, Tom C. T. (2008). Audition. In: *Neuroscience in medicine.* Springer.
35. Kim, N., Homma, K., Puria, S. (2012). Inertial bone conduction: Symmetric and anti-symmetric components. *Journal of the Association for Research in Otolaryngology.* Springer. (A–B) A FE model of the human auditory periphery consisting of middle ear structures and a simplified uncoiled cochlea. A posterior-medial view, (B) anterior-medial view. The air volumes of the ear canal and middle ear cavity are not included in the model. The walls and the scalar fluid in the cochlea have been shown here as partially transparent to allow visualization of the basilar membrane. Figure 1 shows the FE model of the human auditory periphery, consisting of the middle ear and the cochlea.
36. Lewis, E. R., Leverenz, E. L., & Bialek, W. (1985). *The vertebrate inner ear* (p. 190). CRC Press.

37. Rossing, T. D. (1989). *The science of sound* (2nd ed., pp. 69–70) Addison Wesley.
38. Flanagan, J. L. (1972). *Speech analysis, synthesis and perception* (2nd ed.). New York: Springer-Verlag.
39. von Bekesy, G. The pattern of vibrations in the cochlea. In: *Experiments in hearing* (Chap. 11, p. 407). McGraw Hill.
40. Parthasarathi, A. A., Grosh, K., Nuttall, A. L. (2000, January). Three-dimensional numerical modeling for global cochlear dynamics. *Journal of the American Statistical Association, 107*(1), 474.
41. Phi – 1.81649… is a proportion that figures into the progression of growth, flow and proportions throughout nature.
42. The distinctions of "upper" and "lower" are for visualization only, and are accurate only if you remove the cochlea and set it down like a snail shell on a flat surface. The upper chamber is called the "vestibular canal," the lower is called the "tympanic canal."
43. von Bekesy, G. The pattern of vibrations in the cochlea. In: *Experiments in hearing* (Chap. 11, p. 407). McGraw Hill.
44. Buser, P., & Imbert, M. (1992). *Audition* (p. 176). MIT Press.
45. Lewis, E. R., Leverenz, E. L., & Bialek, W. (1985). Active processes and amplification in the inner ear. In: *The vertebrate inner ear* (pp. 214–220). CRC Press. See also: Kennedy, H. J., Crawford, A. C., & Fettiplace, R. (2005, February 24). Force generation by mammalian hair bundles supports a role in cochlear amplification. *Nature, 433*, 680.
46. Duke, T. (2002, May). The power of hearing. In: *Physics World* reviews the work of various researchers indicating a complex electrical amplification system where the "tip links" of the hair cells actually generate electrical potential proportional to deflection on resonance, providing an electrical amplifier to specific acoustical stimulus. "One can think of the cochlea as containing many "voices," each of which is ready to sing along with any incoming sound that falls within its own range of pitch"—amplifying incoming signals.
47. Davis, A. (1962). Advances in the neurophysiology and neuroanatomy of the cochlea. *Journal of the Acoustical Society of America, 34*, 1377.
48. Khanna, S. M., Decraemer, W. F., Hao, L. F. (2003). Motion of organ of Corti in the apical turn of a living guinea pig. *Proceedings of SPIE, 3411*—In: Tomasini, E. P. (Ed.) *The International Society for Optical Engineering Third International Conference on Vibration Measurements by Laser Techniques: Advances and Applications*, June 1998 (pp. 592–600).
49. Kemp, D. T. (1978, November). Stimulated acoustic emissions from within the human auditory system. *Journal of the American Statistical Association, 64*(5), 1386–1391. Kemp is credited with the discovery of otoacoustic emissions generated by mechanisms within the cochlea as a part of our hearing process.
50. Buser, P., & Imbert, M. (1992). *Audition* (p. 145). MIT Press.
51. Pascal, J., Bourgeade, A., Lagier, M., Legros, C. (1998, September). Linear and nonlinear model of the human middle ear. *Journal of the American Statistical Association, 104*(3), 1509 (Pt. 1).
52. Buser, P., & Imbert, M. (1992). *Audition* (pp. 137–139). MIT Press.
53. Ibid. p. 150.
54. Probst, R., Lonsbury-Martin, B. L., & Martin, G. K. (1991). A review of otoacoustic emissions. *Journal of the American Statistical Association, 85*(5), 2027–2067.
55. With the area difference between the ear drum and the oval window of 1:30 translated into sound pressure exposure → 20 log (30) = 29.54 dB.
56. Buser, P., & Imbert, M. (1992). *Audition* (p. 131). MIT Press.
57. Shaw, E. A. G., Teranishi, R. (1968). Sound pressure generated in an external ear replica and real human ears by a near-by point source. *Journal of the American Statistical Association, 44*(1), 240.
58. Wright, D., Hebrank, J. H., & Wilson, B. (1974). Pinna reflections as cues for localization. *Journal of the American Statistical Association, 56*(3), 957.
59. Traditional Chinese medicine includes a study of auricular acupuncture that maps the entire body on to the auricle. The model is derived from superimposing the human embryo over the

auricle as it would be in the womb, with the head aligned over the ear lobe and the arc of the pinna aligned with the curvature of the spine. The rest of the body, from arms to internal organs, is all circumscribed within the auricle envelope. See: Oleson, T. (2002). *Auriculotherapy manual: Chinese and Western systems of ear acupuncture.* Churchill Livingstone.
60. Dirks, D. D., Malmquist, C. W., & Bower, D. R. (1968). *Toward the specifications of a normal bone conduction threshold. Journal of the American Statistical Association, 43,* 1237.
61. Abbott, A. (2002, March 7). Music, maestro, please! *Nature, 417,* 12.
62. von Békésy, G. (1960). *Experiments in hearing* (pp. 259–261). McGraw Hill. Refers to the threshold of fusion where acoustical pulses become fused into a perceived tone. He equates this to the frame rate in cinematography, where the persistence of vision integrates the flashing movie frames into a continuous moving image — also at about 18 frames a second.
63. Buser, P., & Imbert, M. (1992). *Audition* (p. 38). MIT Press.
64. Nordmark, J. O. (1976). Binaural time discrimination. *Journal of the American Statistical Association, 60*(4), 870. See also: Blackmer, D. E. (1999, January). The world beyond 20 kHz. Studio Sound. This does not imply that we distinguish pitch, or even hear tones in the 500 kHz range (or the 1 Hz range) but only that we are affected by acoustical energy in these time domains.

Notes for Chapter 4

1. Margulis, L. (1998). *Symbiotic planet* (p. 3). Basic Books.
2. Morton, E. S., & Page, J. (1992). *Animal talk: Science and the voices of nature* (p. xxvii). Random House.
3. Jewett, D., & Williston, J. (1971). Auditory evoked far fields averaged from the scalp of humans. *Brain, 94,* 681–696. This work opens the door to the now widely used Auditory Brainstem Response (ABR) used in bio-acoustic research.
4. While there is repeatability between AEP individual animals in AEP thresholds, there is yet no correlation between AEP thresholds and behavioral thresholds. Also "Single Cell Neuron Response" thresholds in some animals can be −20 dB below AEP thresholds. (See: Maruska, K. P., Boyle, K. S., Dewan, L. R., & Tricas, T. C. (2006, November). Sound production and hearing ability in the Hawaiian sergeant fish. *The Journal of the Acoustical Society of America, 120*(5), 3103–3104)
5. The closest approximations of sine waves produced by animal vocalizations are found in the echolocation calls of some bats and dolphins — both mammals with cochleae. While not commonly produced by other animal vocalizations, sinusoidal motion is common in natural environments with the mechanical flexing, pulsing and oscillating motions of the physical surroundings excited by wind, currents, waves and gravity. In sum these noises are akin to "gaussian noise" — and in the cases of fish in turbulent waters or insects surrounded by rustling leaves, this noise can be quite loud. Sensitivity to sinusoidal noise would not serve animals submerged in these types of environments.
6. The liability of sinusoidal frequency testing was first identified in the early days of the Acoustical Society in an article by Rogers H. Galt (Galt, R. H. (1929). Methods and apparatus for measuring the noise audiogram. *The Journal of the Acoustical Society of America, 1*(1), 147–157), but was somehow left behind in the advance of science. Caveats about this systemization were published later by Edmund Prince Fowler, stating that "…the hearing mechanism is not just an electrical hookup." (See: Fowler, E. P. (1943). Is the threshold audiogram sufficient for measuring hearing capacity? *The Journal of the Acoustical Society of America, 13*(1), 57–60). Fowler begins his article with the sentence: "From time immemorial hearing acuity has been thought of, and in fact measured, in terms of the distance a sound could be heard." His article infers a spatial relationship to sound that has since been replaced by an "amplitude" relationship to sound as our testing equipment became technologically "refined."

7. Barth, F. G. (1998). Spider vibration sense. In R. R. Hoy, A. N. Popper, & R. R. Fay (Eds.), *Comparative hearing: Insects* (pp. 255–256). Springer. Spider kleptoparasites can forage in a spider's web without alerting the resident using "sinusoidal gait" that the host spider is unable to perceive.

8. Speck-Hergeröder, J., Barth, F.G. (1987). Tuning of vibration sensitive neurons in the central nervous system of a wandering spider. *Cupiennius salei* Keys. *Journal of Comparative Physiology A, 16*, 476–475.

9. In one ABR apparatus for fish, the specimen was harnessed into a jig in a shallow tank. This tank was firmly mounted to a heavy, acoustically isolated granite table to prevent room-borne vibrations from affecting the tank. About 4 ft above the tank was an open frame 12″ diameter speaker cone aiming down at the tank. While the pressure gradient sound field close to the specimen was monitored with a hydrophone to verify the results, many questions arise for me about the physics of this apparatus.

10. van Noordwijk, A. J. (2002). Excuses for avian infidelity. *Nature, 419*, 517.

11. Tavolga, W. N. (1981). *Hearing and sound communication in fishes* (p.574). Springer Verlag.

12. Does the name "Pavlov" ring a bell?

13. Nestler, J. M., Ploskey, G. R., Pickens, J., Menezes, J., & Schilt, C. (1992). Responses of blueback herring to high-frequency sound and implications for reducing entrainment at hydropower dams. *North American Journal of Fish Management, 12*, 667–683.

14. Mann, D. A., Lu, Z., & Popper, A. N. (1997). A clupeid fish can detect ultrasound. *Nature, 389*, 341.

15. Margulis, L. (1998). *Symbiotic planet: A new view of evolution.* Basic Books. Harman, W. W., & Sahtouris, E. (1998). *Biology Revisioned* (pp. 39–51). North Atlantic Books.

16. Lamarck, Jean Baptiste "Philosophie Zoologique" (1809) Lamarck's evolutionary theory held that the development of the embryo reflected the evolution of all animals; that it was the innate quality of animals to "improve" with successive generations, and Mankind sat at the top of this evolutionary progression.

17. Harman, W. W., & Sahtouris, E. (1998). *Biology revisioned* (pp. 39–51). North Atlantic Books.

18. Stevens, S. S., & Warshofsky, F. (1970). *Sound and hearing* (pp. 64–65). Time-Life Books.

19. Ibid. p. 32

20. Lenhardt, M. L. (1995, May). Low-frequency auditory behavior in sea lampreys, primitive fish, and marine turtles. *The Journal of the Acoustical Society of America, 97*(5), 3371.

21. Clack, J. A., Ahlberg, P. E., Finney, S. M., Dominguez Alonzo, P., Robinson, J., & Ketcham, P. A. (2002). A uniquely specialized ear in a very early tetra pod. *Nature, 425*(6953), 65–69 reveal and even more detailed inner ear indicating a form in the skull akin to the "swim-bladder and weberian ossicles" form in fishes adapted to underwater hearing.

22. Stocker, M. (2002). Fish, mollusks and other sea animals' use of sound, and the impact of anthropogenic noise in the marine acoustic environment. *Soundscape Journal of Acoustic Ecology, 3*(2 & 4), 1.

23. Coombs, S., Fay, R. R., & Janssen, J. (1989). Hot-film Anemometry for measuring later line stimuli. *The Journal of the Acoustical Society of America, 85*(5), 2185–2192.

24. Stevens, S. S., & Warshofsky, F. (1965). *Sound and hearing* (p. 66). Time Life Books.

25. Ibid. p. 68.

26. Ibid. p. 67.

27. Siler, W. (1969). Near and farfields in a marine environment. *The Journal of the Acoustical Society of America, 46*(2.2), 483. See also: Hawkins, A. D. (1981). The hearing abilities of fish. In W. N. Tavolga, A. N. Popper, R. R. Fay (Eds.), *Hearing and sound communication of fishes* (p. 109). Springer-Verlag.

28. Engelmann, J., Hanke, W., Mogdans, J., & Bleckmann, H. (2000). Neurobiology: Hydrodynamic stimuli and the fish lateral line. *Nature, 408*, 51–52.

29. Dehnhardt, G., Mauck, B., Hanke, W., & Bleckmann, H. (2001, July 6). Hydrodynamic Trail-Following in Harbor Seals (*Phoca vitulina*). *Science, 293*, 102–104.

30. Whitlow, W. L. Au. (2002, November). Snapping shrimps: The bane of bioacoustics monitoring of coral reefs. *The Journal of the Acoustical Society of America, 112*(5), 2202.

31. Potter, J. R., & Malod, L. (2002, December). Expanding uses of ambient noise for imaging, detection, and communication. Paper delivered at the First Pan-American meeting on Acoustics. *The Journal of the Acoustical Society of America, 112*(5), 2261.

32. Simpson, S. D., Meekan, M. G., Larsen, N. J., McCauley, R. D., & Jeffs, A. (2010). Behavioral plasticity in larval reef fish: orientation is influenced by recent acoustic experiences. *Behavioral Ecology.* doi:10.1093/beheco/arq117

33. Castle, M. J., & Kibblewhite, A. C. (1975). The contribution of the sea urchin to ambient sea noise. *The Journal of the Acoustical Society of America, 58*(S1), 122.

34. Fish, M. P. (1964). Biological sources of sustained ambient noise. In W. N. Tavolga (Ed.), *Marine bio-acoustics* (pp. 175–194). New York: Pergamon.

35. Vermeij, M. J. A., Marhaver, K. L., Huijbers, C. M., Nagelkerken, I., & Simpson, S. D. (2010, May). Coral larvae move toward reef sounds. *PloS One, 5*, 5.

36. Frings, H., & Frings, M. (1967). Underwater sound fields and behavior of marine invertebrates. In W. N. Tavolga (Ed.), *Marine bio-acoustics.* Oxford, UK: Pergammon Press.

37. National Research Council Ocean Studies Board. (2003). *Ocean noise and marine mammals* (p. 87). National Academies Press.

38. Payne, R. (1970). *Songs of the humpback whale* Living Music. Payne's recordings revealed to the world the complex and haunting songs of these animals in what has become the largest selling nature sound recording.

39. Herodotus. *Histories in nine books* First Logos: 1:1–94.

40. There is some overlap, a range dependence on animal size and species, and some exceptions, but the Odontocetes vocalization and echolocation range is dominantly between 1 kHz and 150 kHz, the Mysticetes vocalization range is between 3 Hz and 40 kHz.

41. Secrets of whales' long-distance songs are being unveiled by U.S. Navy's undersea microphones—but sound pollution threatens (2005, February). Cornell News. See also: Clark, C. W., & Ellison, W. T. (2004). Potential use of low-frequency sounds by baleen whales for probing the environment: evidence from models and empirical measurements. In J. Thomas, C. Moss, & M. Vater (Eds.), *Echolocation in bats and dolphins* (p. 564). University of Chicago Press.

42. Frankel, A. S., & Clark, C. W. (2002). Behavioral responses of humpback whales (*Megaptera novaeangliae*) to full-scale ATOC signals. *The Journal of the Acoustical Society of America, 106*(4), 1930.

43. Ketten, D. R., Starr, F. L., Wartzok, D. (1983). *Comparative cochlear morphology in echolocating cetaceans.* Conference proceedings of the 106th meeting of the Acoustical Society. V74:s6.

44. Ketten, D. R. (1995). Estimates of blast injury and acoustic trauma zones for marine mammals from underwater explosions. In R. A. Kastelein, J. A. Thomas, & P. E. Nachtigall (Eds.), *Sensory systems of aquatic mammals* (pp. 393–394). Woerden, Netherlands: De Spil Publishers.

45. The lipids show some remarkable acoustical properties and play a role on both sound perception as well as phonation. See Koopman, H., Budge, S., Ketten, D., Iverson, S. (2003). *The influence of phylogeny, ontogeny and topography on the lipid composition of the mandibular fats of toothed whales: Implications for hearing.* Paper delivered at the Environmental Consequences of Underwater Sound, Office of Naval Research.

46. Norris, K. S. (1974). Sound transmission in the porpoise head. *The Journal of the Acoustical Society of America, 56*(3), 659.

47. Spermaceti oil contained in the sperm whale melon is liquid at 37 °C moving to a solid below 31 °C with associated density changes, and is thermally regulated by the whale while sounding.

48. Aroyan, J. L., Crawford, T. W., Kent, J., & Norris, K. S. (1994). Computer modeling of beam formation in *Delphinas delphis*. *The Journal of the Acoustical Society of America, 92*(5), 2539.

49. Koopman, H., Budge, S., Ketten, D., Iverson, S. (2003). *The influence of phylogeny, ontogeny and topography on the lipid composition of the mandibular fats of toothed whales: Implications for hearing.* Paper delivered at the Environmental Consequences of Underwater Sound conference, May 2003.

50. Alligators' lower jaws are relatively thin and have a hollow channel filled with lipids with a "window" up by the hinge, exposing this lipid mass to the outer skin covering—which looks

like a drum head and may act as a diaphragm. I have not found any reference to this in the current literature. This all suggests an underwater adaptation to sound perception that warrants further investigation.

51. Personal conversation with Dr. Roger Payne.

52. Solntseva, G. N. (1995). The auditory organ of mammals in relation to the acoustic properties of habitat and frequency tuning. In R. A. Kastelein, J. A. Thomas, & P. Nachtigall (Eds.), *Sensory systems of aquatic mammals* (p. 466). Netherlands: De Spil Publishers.

53. Schwartz, D. M. (1996). Snatching scientific secrets from the hippo's gaping jaws. 1996 Smithsonian. A review of the bio-acoustic work of Bill Barlow. The hippo jaw has other similarities with dolphin jaws.

54. National Research Council Ocean Studies Board. (2003). *Ocean noise and marine mammals*. Color Plate 3. National Academies Press.

55. Evans, W. E. (1973). Echolocation of marine delphinids and one species of freshwater dolphin. *The Journal of the Acoustical Society of America, 54*(1), 194.

56. Herman, L. M., Pack, A. A., & Hoffmann-Kuhnt, M. (1998) Seeing through sound: dolphins (Tursiops truncatus) perceive the spatial structure of objects through echolocation. *Journal of comparative psychology, 14*, 292.

57. Buser, P., & Imbert, M. (1992). *Audition* (p. 110.). MIT Press.

58. See: Fay, R. R. (1988). *Hearing in vertebrates: A psychophysics databook* (p. 353). Winnetka, IL: Hill-Fay Associates.

59. According to the staff at the Lindsey Wildlife Museum in Walnut Creek, California, domestic cats bring down over a million migratory birds per year in the San Francisco Bay area alone.

60. Fay, R. R. (1988). *Hearing in vertebrates: A psychophysics databook* (p.347). Winnetka, IL: Hill-Fay Associates. Summary of behavioral audiograms.

61. "Anthelicine"—pertaining to the antihelix of the ear. "Sulcus"—a furrow or groove.

62. I have noticed this sulcus in dogs, but as suggested in the text, it may be vestigial in most breeds.

63. Beitel, R. E. (1996). Acoustic pursuit of invisible moving targets by cats. *The Journal of the Acoustical Society of America, 105*(6), 3449.

64. Echolocation is present in all microchiropterans, but only in one species of megachiropteran—who otherwise use their eyesight. See Kunz, T. H., & Pearson, E. D. (1994). Introduction to *Walkers work of bats* (p. 9) by Ronald M. Nowak. Johns Hopkins University Press.

65. Müller, R. (2004). A numerical study of the role of the tragus in the big brown bat. *The Journal of the Acoustical Society of America, 116*(6), 3701–3712. See also: Gao, L. (2010, March). Ear deformations in the bio-sonar system of the horseshoe bat. *The Journal of the Acoustical Society of America, 127*(3), 2.

66. O'Connell-Rodwell, C. E., Hart, L. A., & Arnason, B. T. (2001). Exploring the potential use of seismic waves as a communication channel by elephants and other large mammals. *American Zoologist, 41*, 1157–1170.

67. Gray, P. M., Krause, B., Atema, J., Payne, R., Krumhansl, C., Baptista, L. (2001, January). The music of nature and the nature of music. *Science, 291*(5), 52–54.

68. Heffner, H., Masterson, B. (1980). Hearing in glires: Domestic rabbit, cotton rat, feral house mouse, and kangaroo rat. *The Journal of the Acoustical Society of America, 68*(8), 1584–1599.

69. Magal, C., Scholler, M., Tautz, J., & Casas, J. (2000). The role of leaf structure in vibration propagation. *The Journal of the Acoustical Society of America, 105*(5 pt. 1), 2412–2418.

70. Michelsen, A., Fink, F., Gogalga, M., & Traue, D. (1982). Plants as transmission channels for insect vibrational songs. *Behavioral Ecology and Sociobiology. 11*, 269–281.

71. Cardoso, A. J., & Heyer, W. R. (1995). Advertisement, aggressive, and possible seismic signals of the frog Leptodactylus syphax. *Alytes, 13*, 67–76.

72. Barnett, K. E., Cocroft, R. B., & Fleishman, L. J. (1999, May). Possible communication by substrate vibration in a chameleon. *Herp Beat, 10*(5).

73. Belwood, J. J., & Morris, G. K. (1987). Bat predation and its influence on calling behavior in neotropical katydids. *Science, 238*, 64–67.

74. Greenfield, M. D. (2002). *Signalers and receivers* (pp. 157–158). Oxford University Press.

75. Kimchi, T., Reshef, M., & Terkel, J. (2005). Evidence for the use of reflected self-generated seismic waves for spatial orientation in a blind subterranean mammal. *The Journal of Experimental Biology, 208*, 647–659.
76. Neural processing may happen in the brain or in distributed ganglion more local to the specific stimulus. See: Pollack, G. S. (1998). Neural processing of acoustic signals. In R. R. Hoy, A. N. Popper, & R. R. Fay (Eds.), *Comparative hearing: Insects* (p.139). Springer.
77. Downer, J. (1988). *Supersense: Perception in the animal world* (p.73). London: BBC Enterprises.
78. Gibson, G., & Russell, I. (2006). Flying in tune: Sexual recognition in mosquitoes. *Current Biology, 16*(13), 1311–1316. In this study mosquitoes of the opposite sex will tailor their wingbeat frequency until their flight tones match. By contrast, same sex pairs diverge in frequency.
79. Downer, J. (1988). *Supersense: Perception in the animal world* (p.73). London: BBC Enterprises.
80. Robert, D., & Hoy, R. R. (1998). Tympanal hearing in diptera. In R. R. Hoy, A. N. Popper, & R. R. Fay (Eds.), *Comparative hearing: Insects* (p. 200). Springer.
81. Miles, R. N., Robert, D., & Hoy, R. R. (1995). Mechanically coupled ears for directional hearing in the parasitoid fly *Ormia ochracea. The Journal of the Acoustical Society of America, 98*(6), 3059–3070.
82. Michelsen, A., & Löhe, G. (1995). Tuned directionality in cricket ears. *Nature, 375*, 639.
83. Farris, H. E., & Hoy, R. R. (2000). Ultrasound sensitivity in the cricket, *Eunemobius carolinus (Gryllidae, Nemobiinae). The Journal of the Acoustical Society of America, 107*(3), 1727–1736.
84. Fullard, J. H. (1998). The sensory coevolution of moths and bats. In R. R. Hoy, A. N. Popper, & R. R. Fay (Eds.), *Comparative hearing: Insects* (pp. 279–326). Springer.
85. Fullard, J. H. (1991, Summer). Predator and Prey: Life and death struggles (BATS, Vol. 9/2). Austin, TX: Bat Conservation International.
86. Forester, L. (1982). Visual communication in jumping spiders (*Salticidae*). In P.N. Witt, & J. S. Rovner (Eds.), *Spider communication* (p.161). Princeton University Press.
87. Barth, F. G. (1998). Spider vibration sense. In Hoy, R. R., Popper, A. N., & Fay, R. R. (Eds.), *Comparative hearing: Insects* (p. 244–249). Springer.
88. Klärner, D., & Barth, F. G. (1982). Vibratory signals and prey capture in orb-weaving spiders. *Journal of Comparative Physiology A, 148*, 445–455.
89. Knudsen, E. I., & Konishi, M. (1979). Mechanisms of sound localization in the barn owl (*Tyto alba*). *Journal of Comparative Physiology A, 133*, 13–21.
90. Saunders, J. C., Kieth Duncan, R., Doan, D. E., & Werner, Y. L. (2000). The middle ears of reptiles and birds. In R. J. Dooling, R. R. Fay, & A. N. Popper (Eds.), *Comparative hearing: Birds and reptiles* (p. 63). Springer-Verlag.
91. Kriethen, M. L., & Quine, D. B. (1979). Infrasound detection by the homing pigeon: A behavioral audiogram. *Journal of Comparative Physiology A, 129*, 1–4.
92. Quine, D. B. (1982). Infrasounds: A potential navigation cue for homing pigeons. In *Avian navigation* (p.373–376). Springer-Verlag.
93. Hagstrum, J. T. (2000). Infrasound and the avian navigational map. *Journal of Experimental Biology, 203*, 1103–1111.

Notes for Chapter 5

1. John T. Emlen, Jr. (1959) in his introduction to *Animal sounds and communication* (p. ix) edited by W.E. Lanyon and W.N. Tavolga. American Institute of Biological Sciences.
2. Shipley, J. T. (1945). *Dictionary of word origins* (p. 187). Philosophical Library. under "*immunity.*"
3. Plato (~395–385 B.C.E.). *Cratylus.* (B. Jowett, Trans.). Oxford University Press.

4. Ibid. pp. 407–408.
5. Morton, E. S., & Page, J. (1992). *Animal talk: Science and the voices of nature* (pp. 24–25). Random House. W.H. Thorpe's (1963) work on this was published in *Learning and instinct in animals*. London: Methuen.
6. Sheppard, P. (1996). *The others: How animals made us human* (pp. 86–87).
7. Lao Tsu. (1972). *Tao Te Ching* (first chapter) (Gia-Fu Feng, Trans.). Random House.
8. See: Aitchison, J. (2000). *The seeds of speech: Language origin and evolution* (p. 13). Cambridge University Press for an array of language originating views.
9. Rousseau, J. J. (1852/1966). On the origins of language. In *On the origin of language: Essays by Rousseau and Herder* (J. H. Moran & A. Gode, Trans.). University of Chicago Press.
10. Feld, S. (1990). *Sound and sentiment: Birds, weeping, poetics, and song in Kaluli expression* (p. 34). University of Pennsylvania Press.
11. Griffin, D. R. (1992). *Animal minds* (pp. 156–160). University of Chicago Press.
12. Ackers, S. H., & Slobodchikoff, C. N. (1999). Communication of stimulus size and shape in alarm calls of Gunnison's prairie dogs. *Ethology, 105*, 149–162.
13. Long, M. E. (1998, April). The vanishing Prairie dog. *National Geographic, 193*(4), 122.
14. Shepard, P. (1996). *The others: how animals made us human* (pp. 15–16). Island Press/Shearwater Books.
15. See: Hauser, M. D., Chomsky, N., & Tecumseh Fitch, W. (2002, November 22). The faculty of language: What is it, who has it, and how did it evolve? (p. 1576). *Science, 298*, 1569–1579.
16. Enggist, P. (1996). Dialects in ravens: New aspects of an old problem. Pfister, U. Communication in ravens: Again new aspects of an old problem. *International Bioacoustics Council. XV Symposium.*
17. Baptista, L. F., & King, J. R. (1980, August). Geographical variation in song and song dialects of Montane white-crowned sparrows. *The Condor, 82*(3), 267–284.
18. Baptista, L. F. (1996). Nature and its nurturing in avian vocal development. In D. E. Kroodsma & E. H. Miller [Eds.], *Ecology and evolution of acoustic communication in birds* (pp. 39–60). Cornell University Press.
19. Deecke, V. B., Slater, P. J. B., & Ford, J. K. B. (2002). Selective habituation shapes acoustic predator recognition in harbor seals. *Nature, 420*, 171–173.
20. Morton, E. S., & Page, J. (1992). *Animal talk: Science and the voices of nature* (pp. 252–253). Random House.
21. Ibid. pp. 199–203.
22. Personal conversation with author See: Jenkins, E. B. (1997). *Initiation: A woman's spiritual adventure in the Andes.* G.P. Putnam.
23. James Kale McNeley. (1981). *Holy wind in Navajo philosophy* (p. 15). University of Arizona Press.
24. Maladoma Patrice Somé. (1999). *The healing wisdom of Africa* (p. 65). Putnam/Tarcher.
25. Davis, W., Mackenzie, I., Kennedy, S. (1995). *Nomads of the dawn* (p. 33). Pomegranate Artbooks San Francisco.
26. James Kale McNeley. (1981). *Holy wind in Navajo philosophy* (. p. 12, 17). University of Arizona Press. The winds of the cardinal directions suspend all that exists within their embrace. Some are not so good, accounting for infection, delusion, and deception. But also "...the surrounding Air, transformed by the Air streaming from the breast of the speaker, becomes the means by which the Holy People are controlled." pp. 57–58.
27. Also known as the "Papago," a descriptive name given to them by others.
28. Abram, D. (1996). *The spell of the sensuous* (pp. 190–192). Vintage. Discusses the Hopi language that is not conjugated in the context of linear time and its consequence three dimensional space that serves as the basis of the English language, rather their language infers an experiential context of a continual "manifesting" of the world.
29. Abram, D. (1996). *Spell of the sensuous* (pp. 154–155). Vintage.
30. Shepard, P. (1996). *The others: How animals made us human* (pp. 284–290). Island Press/Shearwater Books.

31. Levitt, H., & Webster, J. C. (1979). Effects of noise and reverberation on speech. In Harris, C. M. (Ed.), *Handbook of acoustical measurement and noise control* (Chap. 16). McGraw-Hill.

32. Pinker, S. (1995). *The language instinct* (pp. 158–163). HarperPerennial. This book is a good read on the topic of language and sound communication.

33. Shaywitz, B. A., Shaywitz, S. E., Pugh, K. R., Constable R. T., Skudlarski, P., Fulbright, R. K., et al. (1995). Sex differences in the functional organization of the brain for language. *Nature, 373*, 561–562.

34. Murphy, G. (2003). Dyslexia: Lost for words. *Nature, 425*, 340–342.

35. Shaywitz, S. E. (1996). *Dyslexia* (pp. 98–104). Scientific American.

36. Madule, P. (1989). The Dyslexified world. In T. M. Gilmore, P. Madule, & B. Thompson (Eds.), *About the Tomatis method* (pp. 45–61). Listening Center Press.

37. Probably *Nemobius* sp.

38. Hickling, R., & Brown, R. L. (2000). Analysis of acoustic communication by ants. *Journal of the American Statistical Association, 108*(4), 1920–1929.

39. Murray Schafer, R. (1977/1994). *The soundscape: Our Sonic environment and the tuning of the world* (pp. 214–217). Destiny Books.

40. Lemon, R. E., Strurger, J., Lechowicz, M. J., & Norman, R. F. (1981). Song features and singing heights of American warblers: Maximization or optimization of distance. *Journal of the American Statistical Association, 69*(4), 1169. This paper refers to how bird song frequencies match the absorption characteristics of the various forest levels—from leaf-litter floor to the dense leafy canopy, or the branch filled areas between.

41. Krause, B. (1987, Winter). Bioacoustics, Habitat Ambience in ecological balance. *Whole Earth Review, #57*.

42. von Muggenthaler, E. (2000, December 6). *Low frequency and infrasonic vocalizations from tigers*. Paper 3aABb1 presented Wednesday morning. ASA/NOISE-CON 2000 Meeting, Newport Beach, CA.

43. A common anecdotal account.

44. von Muggenthaler, E., Baes, C., Hill, D., Fulk, R., & Lee, A. (1999). *Infrasound and low frequency vocalizations from the giraffe; Helmholtz resonance in biology*. Published in the proceedings of Riverbanks Consortium.

45. Evans, N. (1996). *The horse whisperer* (Fiction). Dell books.

46. von Muggenthaler, E., Hale, P., & Ralph, R. Conti "Infrasound from *Equus Caballus*". *Journal of the American Statistical Association.*

47. Randall, J. A. (1994). Discrimination of foot-drumming signatures by kangaroo rats, *Dipodomys spectabilis*. *Animal Behavior, 47*(1), 45–54.

48. Ref. Woodpeckers won't use just any grub-eaten tree to signal, rather they will seek trees with a particular sound suited to their tastes.

49. Dantzker, M. S., Deane, G. B., & Bradbury, J. W. (1999). Directional acoustic radiation in the strut display of male sage grouse *Centrocercus urophasianus*. *Journal of Experimental Biology, 202*, 2893–2909.

50. Fine, M. L., Schrinel, J., & Cameron, T. M. (2004). The effect of loading on disturbance sounds of the Atlantic croaker Micropogonius undulatus: Air versus water. *Journal of the Acoustical Society of America, 116*(2), 1271–1275.

51. de Boer, B. (2009). Acoustic analysis of primate air sacs and their effect on vocalization. *Journal of the American Statistical Association, 126*, 6.

52. Benzon, W. L. (2001). *Beethoven's Anvil: Music in mind and culture* (p. 6). Basic Books.

53. Hingee, M., & Magrath, R. D. (2009, December 7). Flights of fear: a mechanical wing whistle sounds the alarm in a flocking bird. *Proceedings ofthe Royal Society B: Biological Sciences, 276*(1676), 4173–4179. The authors examine the alarm responses to the wing whistling produced by the eighth primary feathers of the crested pigeon *Ocyphaps lophotes*. The amplitude and other characteristics of this whistle vary depending on the speed and force of the wing beats. It follows that a startled bird would be more forceful in its rapid retreat from a threat than if it were just departing into flight. The sound characteristics of the whistling cue conspecifics within the flock of an alarm situation.

54. daVinci, L. *"The notebooks of Leonardo daVinci"* arranged and rendered into English (p. 268). New York: George Braziller.
55. Galambos, R. (1942, Autumn). The avoidance of obstacles by flying bats: Spallanzani's ideas (1794) and later theories. *Isis, 34*(2). Published by: The University of Chicago Press on behalf of The History of Science Society.
56. Hartridge, H. (1920, August 19). The avoidance of objects by bats in their flight. *Journal of Physiology, 54*(1–2), 54–57.
57. Griffin, D. R., & Galambos, R. (1941). The sensory basis of obstacle avoidance by flying bats. *Journal of Experimental Zoology,* 86, 481–506.
58. Gross, C. G. (2005). *Donald R. Griffin, 1915—2003: A biographical memoir. Biographical Memoirs* (Vol. 86). The National Academies Press.
59. Ibid.
60. Epifanio, C. L., Potter, J. R., Deane, G. B., Readhead, M. L., & Buckingham, M. J. (1999). Imaging in the ocean with ambient noise: the ORB experiments. *Journal of the Acoustical Society of America,* 106(6), 3211–3225.
61. Fullard, J. H. (1998). The sensory coevolution of moths and bats. In R. R. Hoy, A. N. Popper, & R. R. Fay (Eds.), *Comparative hearing: Insects* (pp. 279–326). Springer.
62. Mann, D. A., Lu, Z., & Popper, A. N. (1997). A clupeid fish can detect ultrasound. *Nature, 389,* 341.
63. Potter, J. R., & Malod, L. (2002, December). Expanding uses of ambient noise for imaging, detection, and communication. Paper delivered at the First Pan-American Meeting on Acoustics. *Journal of the American Statistical Association, 112*(5), 2261.
64. The determination of the longer extents of time domain perception varies greatly. Musicians may be able to integrate sequential time events of a few seconds to many minutes—allowing them to understand the larger lyrical arcs of music; meanwhile some people are so arrhythmic that they cannot synchronize to any musical beat. See Benzon, W. (2001). Computations and timing in a nervous system. In *Beethoven's Anvil: Music in Mind and Culture* (pp. 57–59). Basic Books.
65. Russell, S. A. (2003). *An obsession with butterflies: Our long love affair with a singular insect* (p. 69). Perseus Publishing.
66. Alexander, R. D. (1957, March). Sound production and associated behavior in insects. *The Ohio Journal of Science, 57*(2), 101–113.
67. Yack, J. E., Smith, M. L., & Weatherhead, P. J. (2001). Caterpillar talk: Acoustically mediated territoriality in larval Lepidoptera. *Proceedings of the National Academy of Sciences, 98*(20), 11371–11375.
68. Dethier, V. G. (1992). *Crickets and katydids, concerts and solos* (pp 22–29). Harvard University Press.
69. Carl Gerhardt, H., & Huber, F. (2002). *Acoustic communication in insects and anurans: Common problems and diverse solutions* (Chaps. 8–10). University of Chicago Press.
70. I used a phase matched pair of Earthworks QTC1 omni directional microphones, ±1 dB from 4 Hz to 40 kHz, through a Mackie VLZ mic preamp +0 dB/−1 dB from 20 Hz to 60 kHz into a Tektronix 2445 150 MHz Oscilloscope.
71. Greenfield, M. D. (2002). *Signalers and receivers: Mechanisms and evolution of arthropod communication* (pp. 150–154). Oxford University Press.
72. Michesen, A., & Löhe, G. (1995). Tuned directionality in cricket ears. *Nature, 375,* 639. See also: Carl Gerhardt, H., & Huber, F. (2002). *Acoustic communication in insects and anurans: Common problems and diverse solutions* (Chapter 7 on Sound localization). University of Chicago Press.
73. Carl Gerhardt, H., & Huber, F. (2002). *Acoustic communication in insects and anurans: Common problems and diverse solutions* (p. 21). University of Chicago Press.
74. Forrest, T., Ariaratnam, J., & Strogatz, S. (1998). *Synchrony in cricket calling songs: Models of coupled biological oscillators.* Paper presented at the 135th Meeting of the Acoustical Society of America. Most of this work has been done on a medium-scale time domain basis for pulse synchronization, and not in the fine scale cycle synchronization I found in the cicadas.

75. The cicada's fundamental pulse rate of 14 Hz yields a wavelength of ~80 ft. I suspect that these animals may have been close to 80′ apart. If I was 40′ from each, their phase locked signal (displaced in space by one cycle from each other), each of their pulse streams would reach me at the same time. The layout of the trees in my yard supports these dimensions.

76. Poulet, J. F. A., & Hedwig, B. (2002). A corollary discharge maintains auditory sensitivity during sound production. *Nature 418*, 872–876.

77. Wood, D., & Fels, J. (1992). *The power of maps* (p. 17–22). Guilford Press.

78. Knudsen, E. I., & Brainard, M. (1991). Visual Instruction of the Neural Map of Auditory Space in the Developing Optic Tectum. *Science, 253*, 85–87.

79. Orban, G. A., Van Calenbergh, F., De Bruyn, B., & Maes, H. (1985). Velocity discrimination in central and peripheral visual field. *Journal of the Optical Society of America A: Optics, Image Science, and Vision, 2*(11), 1836–1847.

80. Bregman, A. S. (1990). *Auditory scene analysis: The perceptual organization of sound.* MIT Press. A thorough examination of the various ways we reconcile, sort and file our auditory inputs to create a cohesive "auditory scene."

81. Various Artists "Voices Of The Rainforest: A Day In The Life Of The Kaluli People" recorded by Steven Feld 1991 Rykodisk.

82. Chatwin, B. (1987). *Songlines* (p. 292). New York: Penguin Books.

83. Ibid. p. 283.

84. Ibid. p. 108.

85. Hagstrum J. T. (2000). Infrasound and the avian navigational map. *Journal of Experimental Biology, 203*, 1103–1111.

86. Xitco, M. J. Jr., & Roitblat, H. L. (1996). Object recognition through eavesdropping: Passive echolocation in bottlenose dolphins. *Animal Learning & Behavior, 24*(4), 355–365. For a comprehensive literature review on the eavesdropping hypotheses see: Gregg, J. D., Dudzinski, K. M., & Smith, H. V. (2007). Do dolphins eavesdrop on the echolocation signals of conspecifics? *International Journal of Comparative Psychology, 20*, 65–88.

87. Bunnell, S. (1974). The evolution of Cetacean intelligence. In J. McIntyre (Ed.), *Mind in the waters* (p. 58). Scribners/Sierra Club books.

88. Lilly, J. C. (1967). *Lilly on dolphins* (pp. 302–327). Anchor/Doubleday (Originally from "The Mind of the Dolphin" 1967, Chap. 12).

89. Lilly, J. C. (1978). Communication between man and dolphin: The possibilities of talking within other species (p. 156, 158). Julian Press. Dr. Lilly seemed preoccupied with the possibility that "delphinese" was a symbolic language, and that even sonic images had "meaning" as opposed to being just experience. His thoughts on interspecies communication included a "television" that would translate sonifications into visual images that both humans and dolphins could see.

Thanks to Andy Lovas for his illustrations, and Barbara Geisler for her graphics guidance. And a big thanks to Rachel Nishan of TwinOaks.org for proofing and indexing.

About the Author

Michael Stocker is a technical generalist by disposition, a bioacoustician by trade, and a musician by avocation. He is the founding director of Ocean Conservation Research, (www.OCR.org) a scientific research and policy development organization focused on understanding the impacts of and finding solutions to the growing problem of human-generated ocean noise pollution. He currently lives in Marin County, California.

M. Stocker, *Hear Where We Are: Sound, Ecology, and Sense of Place,*
DOI 10.1007/978-1-4614-7285-8, © Michael Stocker 2013

Index

A

Aboriginal Australians, 32–33, 143, 161–162
Abram, David, 35, 144
Absorptivity (sound), 19, 52, 71, 74–77,
    182n40. *See also* Reflectivity
    (sound)
  in anechoic chamber, 100–101
  of environmental habitats, 150, 155
  interaural time difference and, 78, 100
  ultrasonic communication and, 154
  in womb, 8
Absurd, etymology of word, 40
Accents/dialects
  of humans, 57, 142
  of nonhuman animals, 141–142
Acoustical cohesion, 150–153, 157–159
Acoustical daylight, 111, 155, 162
Acoustical energy, x, xi, 67–101. *See also*
    Hearing, science of (humans);
    Hearing, science of (nonhumans);
    Thresholds/sensitivity range; Time
    domain perception
  aquatic animals' processing of, 110, 116
  bodily processing of, 99
  common misunderstandings of physics of,
    68–69
  definition of, 3, 67
  extent of human perception of, 76, 100,
    105, 150, 176n64
  health/healing and, 65
  HIDAs and, 60
  infrasonic, 131, 150–153, 156
  nonhuman animals' processing of, 110,
    112, 113, 133, 150, 152, 155
  physical mechanics of, 68–70, 72, 73,
    75, 77
  transmission of by air, 61, 67, 70–72,
    77, 121
  transmission of by earth, 121–124

transmission of by other materials/
    substances, 71, 73
transmission of by water, 60–61, 64–65,
    73, 108, 110, 112, 113, 155
ultrasonic, 127, 150, 153–156, 163
Acoustic communities, 27–30, 107, 148–150.
    See also Soundscapes
  bells and, 41–45
  of religion, 41–47
"Acoustic Harassment Devices," 59
Acoustic shadows, 78
Adaptation of humans
  bioacoustic, 103
  ear anatomy and, 91
  gender-based perception and, 4–5
  linguistic skill and, 35
Adaptation of nonhuman animals, 133,
    178n50. *See also* Bioacoustic
    adaptation
  acoustical niches and, 149
  of arthropods, 124
  arthropods and, 124–125, 129
  behavioral, 118
  bioacoustic, 104, 107, 113, 119, 149,
    150, 155
  bioacoustic niches and, 148–150
  of birds, 129–130, 152
  bulla as mobile or fused and, 114
  cooperative, 107, 113, 157
  of domesticated dogs, 116
  ear anatomy and, 118, 119
  of owls, 129–130
  "secondary thresholds" and, 106
  tremulation and, 122–123
Africa, 44–45
Air, acoustical energy in, 61, 67, 70–72, 121
  reverberation and, 77
Aircraft, 60
A is for Ox (Sanders), 39–40

Made in the USA
Middletown, DE
22 February 2019